The Word & The Spirit:

How God Speaks To You

By Charles H. Huettner

ISBN 978-1-62141-688-3

Printed in the United States of America.

Booklocker.com, Inc.
2012

Acknowledgements

Imagine what it is like to feel the presence of God and the Holy Spirit in your life and not know how to tell others about the joy and comfort that comes from Them each day. Imagine reading the Bible each night and knowing that it is the way to enter into a dialogue with God yet not know much about what is behind and between the words in the text. The answers came together for me during my courses at the Virginia Theological Seminary, especially those conducted by Reverend Doctor Roger Ferlo[1]. Dr. Ferlo's courses in "How to Read the Bible" provided me with the opportunity to learn the big picture of what the ancient interpreters, modern scholars, Christian theologians, and Christians around the world have discovered as they search for the truth about God and how He speaks to us. Dr. Ferlo has a real talent to convey in interactive two-hour sessions the essence of material that scholars have assembled over hundreds of years and seminaries teach to aspiring priests and ministers. It was an extraordinary experience to be one of his students and I thank him for inspiring me to write this book.

Thanks too to Dr. Patricia Lyons. Her comments and course at Virginia Theological Seminary helped me to crystallize my world-view and add richness to this book.

I also thank my friends who come from diverse religious backgrounds for their insights and comments as they reviewed drafts of this book.

---

[1] The Reverend Dr. Roger A. Ferlo, Ph.D., D.D. is the President, Bexley Hall and Seabury- Western Seminary Federation. He was previously Associate Dean and Director of the Institute for Christian Formation and Leadership, and professor of religion and culture at the Virginia Theological Seminary. He is a graduate of Colgate University, where he now serves as a trustee. He completed his Ph.D. at Yale.

Sarah Becker
Mary Elizabeth Beim
Joe Estafen
Richard Geissler
Herb Kaiser
Lee Olsen
Charles Queener
Liz Ropers
Larry Shirk
Charlie Walker

Finally, I thank Jesus Christ and the Holy Spirit for loving us, saving us, and revealing so much to us all.

About the Author

Like Jesus' followers during his days on earth, Charles has been drawn into the ministry from another profession. He is the President of Charles Huettner Associates LLC, an aerospace consulting firm. In addition, he is the Executive Director of the Aerospace States Association (ASA) that is comprised of Lt. Governors and delegates from the nation's 50 states focusing on aerospace related education and economic development issues. Charles has a talent for understanding the "Big Picture". He has had the opportunity to use that talent in his professional career and his Bible study.

Charles retired after 33 years of government service as the Executive Director of the Presidential "Commission on the Future of the U.S. Aerospace Industry". Prior to this, he was the Senior Policy Advisor for Aviation for the National Science and Technology Council (NSTC), in the Executive Office of the President under the Clinton and G.W. Bush Administrations. He was also the Director for Aviation Safety Research at NASA. At FAA, he rose through the ranks from inspector to serve as Deputy Associate Administrator for Aviation Safety. In 2005, he published the National Institute of Aerospace report "Responding to the Call: Aviation Plan for American Leadership.

Charles is an Airline Transport Pilot rated in the B-747, B- 727 and the Air Force C-141 Starlifter. He retired as a Colonel in the USAF Reserves where he last served as the Reserve Augmentee to the Air Force Chief of Safety. His decorations include the Legion of Merit, Meritorious Service Medal, the Air Medal, the Air Force Commendation Medal, the Armed Forces Expeditionary Medal, the Combat Readiness Medal, and the Vietnam Service Medal.

Charles has a BS Degree in Industrial Management from the University of Akron and a Masters Degree in Public Administration from Harvard's John F. Kennedy School of Government. He has a Diploma in Theological Studies from the Virginia Theological Seminary.

His first book, "Jesus Reveals Revelation" provides spiritual insights into God's prophecy to us in the last book of the Bible, "The Revelation to John".

The Word & The Spirit:

How God Speaks To You

By Charles H. Huettner

The Word & The Spirit:
How God Speaks To You

Table of Contents

Introduction: 1
Chapter 1. God's Communication Model 5
Chapter 2. God 11
Chapter 3. Ancient Interpreters 23
Chapter 4. Scholars 35
Chapter 5. Paul 51
Chapter 6. The Church 71
Chapter 7. John 83
Chapter 8. Revelation 95
Chapter 9. Context 111
Chapter 10. Listening to God 127
Epilogue 135
Appendix 1: God's Communications Model 143
Appendix 2: Another Way God Speaks To Us – Mathematics 145

The Word & The Spirit:
How God Speaks To You

Introduction:

If you are reading this, then God is calling you. His spirit is drawing you toward Him and this book. I offer you a fresh and inspired way to learn how to hear what God wants to say to you.

Are you someone who only believes in what you can see and touch; yet you have a feeling that there is something more. You might love your church and attend it regularly, but want to learn more than what you have in worship service. Perhaps you have never actually read the Bible and don't know where to begin. You might have had a bad experience in church or have been hurt by people who believe that they are acting in the name of religion. As a result, you may have turned away from the church, yet you still seek answers. That is what this book will provide. It will describe how God will speak to you. It's not about you and the church, its about you and God.

If we cut through all that is said about God, religion, and the Bible there is a central message that God has told eyewitnesses to tell us. This message sometimes gets overlooked in all the religious discussion. This book will provide you a big picture view that ties together scholarly work, Christian belief, and the Spirit's work in your life. The ideas presented here offer you a different frame of reference for your belief.

This book will invite you to engage with the Bible and God's Spirit real time. It will offer you an opportunity for greater intellectual thought and spiritual feelings. God also speaks to us through mathematics including relativity theory, quantum mechanics, and string theory. We will explore a common thread between Bible scholarship and scientific discovery.

God does care about you, and wants to have a personal relationship with you. Once you see the big picture, you begin to see that God is greater than we have imagined and come to understand the power of the Bible and the Holy Spirit in your life. The result for you will be a more joyful daily life, less fear of death, and strength to take action in the face of difficulties and opportunities.

In the development of this book I was inspired by a chart from an article by Carlos Mesters[2], *Listen to What the Spirit is Saying to the Churches*. In this article Dr. Mesters uses a chart to illustrate his ideas of "listening to God today" from his work with poor Christian communities in Brazil. Dr. Mesters' Brazilian example of "Popular Interpretation" of the Bible makes the case that poor oppressed people can see the Bible as "a mirror or 'symbol' of what they live today". ' If God was with those people then, in the past, then he is also with us now in this struggle we are waging to free ourselves.' They discover that God-with-us is the heart of Scripture. Once this occurs, communities of believers read and interpret the Bible based on their situation and understanding. "The aim of interpretation is no longer to interpret the Bible, but to interpret life with the help of the Bible." The results in the community are actions that are focused on service and, sometimes, political action. Service focuses the community back on the Scripture to gain further knowledge, which brings the community closer, leading to greater service as the circular process continues. The result of this process is how the community listens to God today. This is happening in over 6000 communities in Brazil.

---

[2] *Article first appeared in Concilium 1(1991), pp. 100-11.* The translator is Francis McDonagh.

The first thing that struck me in reading Dr. Mester's essay was the similarities to the early churches that Paul founded. While the poor in Brazil see the Bible stories resembling them, I was seeing how the Brazilian stories resemble the early church. In the early church, people began meeting in homes, read Old Testament (OT) texts, and received the gospel of Christ by word of mouth, then through letters from Paul. The early believers became communities of believers and provided service to each other. Those early communities of believers formed the basis for church belief today. It appears that God's methods of speaking to us today are the same as they have been in the ancient past. Only the context is different.

This brings up the issue of what the Bible really is and how God uses it and the Holy Spirit to speak to each of us today. This is the basis of this book. Like Dr. Mester, I will use a chart that provides a model of how God communicates with us. The chart will be our big picture guide for this book.

Chapter 1. God's Communication Model

John 6:44 (NASB)
44 "No one can come to Me unless the Father who sent Me draws him; and I will raise him up on the last day.

Growing up, I had a simplistic idea of God and the Bible. I thought that:

- the authors of the Bible were inspired by God to sit and write the words that have been translated into what we now read.
- the New Testament (NT) Gospels were written by eye witnesses to Jesus' life and resurrection,
- the Council of Nicea canonized the Bible as it is today,
- the Bible contains the single historical storyline that we are to believe, and
- that the scholars who discounted that story line were distracting us from the truth and our faith.

Almost all of this is wrong. The facts show that the three thousand year story of the Bible is much more complex. The surprise is that this complexity shows that God is greater than what we have imagined Him to be.

To understand how God speaks to you, we need to understand how He has spoken to our ancestors for thousands of years. Through study and awareness, we can come to a deeper understanding of the Bible and the work of God's Spirit in our lives today. I have developed the following model to help me tell the story and for you to use in your own efforts to listen to God. As we progress through this book, we will fill in the details behind this simple model. A copy of this chart is in Appendix 1 to which you may refer to as you read this book.

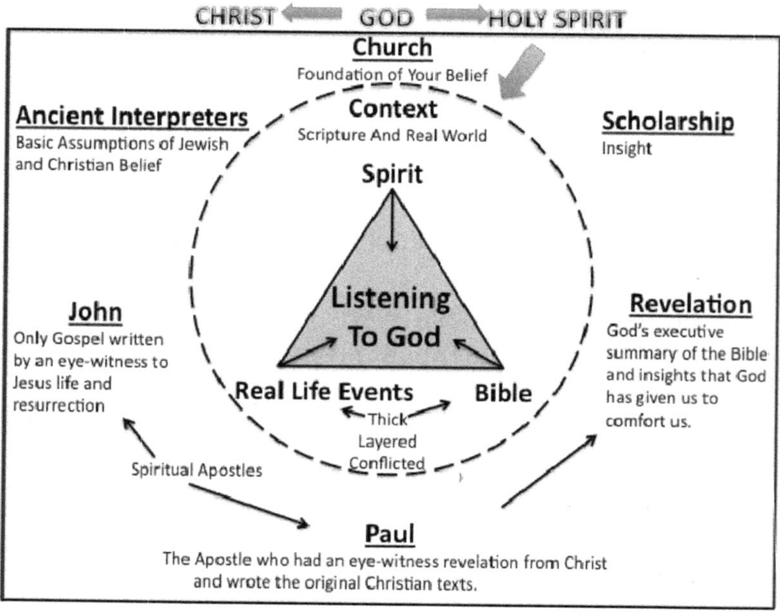

I will briefly summarize the meaning of this chart. We will then devote a chapter to each section to provide the details. It would take a library to fully explain the concepts that are expressed in this chart. In fact, scholars have devoted their entire lives to studying individual aspects of what is captured here. My hope is that the chart will serve as a framework for your relationship with God. I will provide references to books that will help you dig deeper as we progress.

An adaptation of Dr. Mester's chart is the triangle in the center of this chart. His theme is how the communities in Brazil listen to God today. The Bible, the Holy Spirit, and the real world come together for us to listen to God today. God speaks to our heart, mind, and spirit. The key to understanding more about God and the Holy Spirit is to learn more about the Bible and its message to us. That is where the other parts of this chart become important.

Surrounding the triangle in the center is the circle labeled context. There are actually two contexts involved. The first is the context of the passages before and after the scripture being focused on, and other parts of the Bible. The second is the background of the person or community searching for insights from the Bible. Both lead to greater insight into self and God's word, and are the field within which the Holy Spirit operates. I have also included Dr. Walter Brueggemann's concept of the three dimensions of both self and of the Biblical text - thick, layered, and conflicted. I will expand on this later.

Outside the circle are the primary forces that influence the Biblical context. We begin in the upper left corner with the Ancient Interpreters. There are four basic assumptions that form the foundation of what we believe today. From these interpretations, we move to the foundation of the Church. It began in people's homes. The early churches established the original interpretations of the Bible. When Constantine adopted Christianity, the church turned into the imperial Catholic Church with the trappings of imperial splendor. After the reformation, we see many denominations emerge based on their community of interpretation and belief. In some respect, it seems the church is coming full circle with scholars recognizing communities of belief reflecting early Christian churches. Church doctrine and theology have played and continue to play a large role in the context of how we read and understand the Bible. For most people, the church within which they were raised formed their basic belief in God.

The next place around the circle is Biblical scholarship. Scholarship adds insight and historical context to understanding the Bible. While scholar's search for many years has been to find a definitive answer to questions posed by Biblical text, it has actually shown that no single answer exists. In fact, Biblical scholarship has shown the Bible to be like the atom[3], the deeper you look the more you find that there is no mass there, only energy and a probability of mass. Perhaps that says something about God. Yet scholarly insights do help us to understand the Bible better. For example, one of the lessons from scholarly research on the Apostle Paul is that Paul probably only wrote seven letters and that these letters are the oldest existing Christian texts. That brings us to the bottom of the circle on the graph.

Christ's revelation to Paul was also His revelation to us. We will devote an entire chapter to Paul.

The other spiritual apostle is John. The book of John has in it the basic information we need to believe in Jesus Christ. It says so itself.

Paul and John are the only authors in the Bible that Jesus spoke to directly.

But, there are many questions unanswered. That is where research into the book of Revelation comes in. The Holy Spirit worked to ensure that Revelation was written, made part of the canon, and not modified over the centuries for a reason. It contains God's executive summary of the Bible and insights that God has given us to comfort us and give us assurance during difficult times including the end-times as you will see.

Is it any wonder that man has not come to a definitive view of God's word with the combination of so many variables and contexts? I hope to show you that it is because of these variables that God is able to transform each of us according to His will.

---

[3] See Appendix 2.

This is a very brief overview of the chart from the inside out. We will begin our journey of understanding in reverse, from the outside in.

Let's begin with God.

Chapter 2. God

The first two questions one might ask is why should we believe in the God of the Jews and Christians, and why should we believe that the Bible is the book that God uses to communicate with us? Clearly there have been thousands of civilizations since the beginning of mankind and there have been thousands of gods worshiped by these civilizations. Many had sacred stories passed down by word of mouth. When writing became possible they created sacred texts. What distinguishes our God and our God's Bible from all other religions is that He is a God that has sacrificed Himself for us, speaks to the heart and mind of individuals, and wants a personal relationship with each one of us.

Jesus made unique claims about himself that requires us to believe that he is the Son of God or a liar. One cannot simply think of him as a great moral teacher.

John 1:14 (NASB)
14 And the Word became flesh, and dwelt among us, and we saw His glory, glory as of the only begotten from the Father, full of grace and truth.

What God begets is God. What God creates is not.

John 4:25-26 (NASB)
25 The woman said to Him, "I know that Messiah is coming (He who is called Christ); when that One comes, He will declare all things to us."
26 Jesus *said to her, "I who speak to you am He."

John 5:21-23 (NASB)
21 "For just as the Father raises the dead and gives them life, even so the Son also gives life to whom He wishes.

22 "For not even the Father judges anyone, but He has given all judgment to the Son,
23 so that all will honor the Son even as they honor the Father. He who does not honor the Son does not honor the Father who sent Him.

John 20:21-22 (NASB)
21 So Jesus said to them again, "Peace be with you; as the Father has sent Me, I also send you."
22 And when He had said this, He breathed on them and *said to them, "Receive the Holy Spirit.

In *The Everlasting Man,* G.K. Chesterton[4] points out that no great moral teacher ever claimed to be God – Not Mohammed, not Confucius, or Plato, or Moses, or Buddha: " Not one of them ever made that claim…and the greater the man is, the less likely he is to make the very grandest claim."

Yet Jesus did make these claims, and many of his followers sacrificed their lives because they knew that His claims were true. Jesus, the son of God, came as a person to teach and save individuals not nations as the Jews thought He would.

Christian scholar, Dr. Dairmaid MacCulloch, makes an important observation about God's relationship with His people. He observes that in the Old Testament (OT), Israel's God is referred to as Father from the very beginning. In the New Testament (NT), Jesus refers to God as "abba," that is an Aramaic word equivalent to "Dad". Abba had never been used to address God before in Jewish tradition. He goes on to say, "There is further proof that this notion of an intimate Fatherhood between God and humanity is a basic layer of Jesus' message: he goes beyond self-reference. In 'The Lord's Prayer", which lies at the

[4]G.K. Chesterton, *The Everlasting Man*, Garden City, N.Y.: Image Books, 1955, p. 201

heart of Christian approaches to God, He tells His disciples to pray to their Father in Heaven- though the followers address God not as abba but by the ordinary Greek word for 'father" pater."[5]

God's relationship with individuals has been documented in an unbroken line from ancient times to the present day. We can trace the story of God's relationship with a multitude of men and women from before written text to the present day in the Christian Bible. This intimate one-on-one relationship is a prerequisite if we are to believe that there is a living God who loves us and cares for each of us. If God wants to speak to us he would find a way to speak to each generation.

Our God has.

But why focus on Abraham, Isaac, and Jacob the predecessors of today's Jews? Because God told us He has. When Moses asked God who He is, God said He is their God not that they are his people.

> "Exodus 3:13-15 (NASB)[6]
> 13 Then Moses said to God, "Behold, I am going to the sons of Israel, and I will say to them, 'The God of your fathers has sent me to you.' Now they may say to me, 'What is His name?' What shall I say to them?"
> 14 God said to Moses, "I AM WHO I AM"; and He said, "Thus you shall say to the sons of Israel, 'I AM has sent me to you.'"
> 15 God, furthermore, said to Moses, "Thus you shall say to the sons of Israel, 'The LORD, the God of your fathers,

---

[5] *Christianity The First Three Thousand Years*, Diarmaid MacCulloch, 2009, Viking Penguin Page 81.

[6] "Scripture taken from the NEW AMERICAN STANDARD BIBLE®, Copyright © 1960,1962,1963,1968,1971,1972,1973,1975,1977,1995 by The Lockman Foundation. Used by permission."

> the God of Abraham, the God of Isaac, and the God of Jacob, has sent me to you.' This is My name forever, and this is My memorial-name to all generations.

This does not mean that God only loves the Jews. No. What it does mean is that He is a God that has a personal relationship with individuals, and that He used the Jews as messengers and preservers of the sacred texts until Jesus Christ came. God has always existed and His presence can be seen in his creation[7]. However, if He was to have a personal relationship with individuals, he had to choose one to begin with. He chose Abraham. God's involvement with the Jews is especially obvious when we consider that they were not a world-dominating culture in ancient times. Many civilizations have worked to eradicate the Jews over the centuries, yet they still prosper and preserve the sacred text to this day.

God has made it so.

But what proof do we have that God is a personal God who speaks to individuals? Here is some supporting evidence (albeit not a comprehensive list) that God works with us one-on-one in the real world.

1. Gen 3:8 (NASB) "They heard the sound of the Lord God walking in the garden". The Bible begins with a description of creating individuals and then describes individual interactions between Himself, Adam, and Eve.

---

[7] Romans 1:20-21 (NASB)
20 For since the creation of the world His invisible attributes, His eternal power and divine nature, have been clearly seen, being understood through what has been made, so that they are without excuse.
21 For even though they knew God, they did not honor Him as God or give thanks, but they became futile in their speculations, and their foolish heart was darkened.

2. Genesis 32:22-30 describes Jacob personally wrestling with God.
3. Exodus 3:16 (NASB) says " The Lord, the God of Abraham, of Isaac, and of Jacob". God describes himself as the God of specific people, not the God of the universe or of nations, or of a specific church, etc.
4. Thousands of additional Bible stories of God's interaction with specific people in their lives including Moses, David, and the Prophets.
5. Jesus was born of one woman, Mary, and entered the world as one person, Jesus, at a specific time in history.
6. Jesus chooses 12 disciples and works one-on-one with all the people he meets teaching some, forgiving some, healing some, and rebuking some.
7. Jesus spoke in parables so all could hear, but he explained the meaning to individual disciples.
8. John 20:11-18 Jesus first reveals himself to one woman after his resurrection Mary Magdalene.
9. John 20:19-23 Jesus appears to his disciples and gives **each** of them the Holy Spirit.
10. John 20:24-29 Jesus appears to Thomas in the disciples' company completing the story of his work on earth by showing us how important it is to Him to respond to each one of us.
11. Acts 2:38 (NASB) 'Peter said to them, "Repent, and **each of you** be baptized in the name of Jesus Christ for the forgiveness of your sins; and **you** will receive the gift of the Holy Spirit.'

Do you see the connection to one-on-one? This is not a god who shouts from the heavens to the people or a god that needs to be constantly appeased out of fear. He is a God who speaks to and through people. God spoke to Moses and then Moses spoke to the people. God did not shout to the people from the mountaintop. God has given us the Holy Spirit as His connection to each of us as described in John 20:22.

My conclusion is that our God is **our** God each one of us: Abraham, Isaac, Rosemary, Mike, Nicole, Yoko, Juan, Hassan, and you. He is certainly the God of the universe, and is all-powerful. However, He chooses to have His name and His story told through individuals in the book that He has inspired. He is still talking to us today through the Bible and His Spirit.

A perfect example of this comes from an interview with Dr. Francis Collins, the director of the National Human Genome Research Institute and leader of the effort to decode human DNA.[8]

> "I finished up my graduate degree in quantum mechanics, but underwent a bit of a personal crisis, recognizing that I didn't want to do that for the rest of my life. It was too abstract, too far removed from human concerns. And so in considerable disarray as far as my own intentions of what I would want to do with the rest of my life, I decided to go to medical school as a way of trying to explore this more human side of science, namely biology.
>
> So it was really as a medical student, and later as a resident, encountering the realities of what disease and the specter of death does to human beings, that I began to wonder about this. Some of my patients were clearly relying very heavily on their faith as a source of strength in circumstances that were pretty awful. They had terrible diseases from which they were probably not going to escape, and yet instead of railing at God, they seemed to lean on their faith as a source of great comfort and reassurance. They weren't somehow, perceiving it as the

---

[8] PBS, *The Question of God, Other Voices, An Interview with Francis Collins,* Copyright © 2004 WGBH Educational Foundation (http://www.pbs.org/wgbh/questionofgod/voices/collins.html)

really awful thing that it seemed to me to be. And that was interesting and puzzling and unsettling.

As I began to ask a few questions of those people, I realized something very fundamental: I had made a decision to reject any faith view of the world without ever really knowing what it was that I had rejected. And that worried me. As a scientist, you're not supposed to make decisions without the data. It was pretty clear I hadn't done any data collecting here about what these faiths stood for.

Now, I was still pretty sure that faith traditions were all superstitions and something that would not apply to me, and something that I wouldn't be interested in. But I did feel compelled to find out a bit more about what it was that I had rejected. So with an intention of shooting this all down, I went to speak to a Methodist minister in Chapel Hill, which is where I was at the time. I sat in his office and made all sorts of accusations, and probably said blasphemous things about the faith that he stood for, but sincerely asked him to help me find out what it was all about. And he was very tolerant and patient and listened and suggested that, for starters, it might be good if I read a little bit more about what these faiths stood for. And perhaps the Bible would be a good place to start. I wasn't so interested in that at that point. But he also said, " You know, your story reminds me a little bit of somebody else who has written about his experience- that Oxford scholar, C.S. Lewis."

I had no idea, really, who Lewis was. The idea that he was a scholar, though, that appealed to my intellectual pride. Maybe somebody with that kind of title would be able to write something that I could understand and appreciate.

So this wonderful minister gave me his own copy of *Mere Christianity*[9], Lewis's slim tome that outlines the arguments leading to his conclusion that God is not only a possibility, but a plausibility. That the rational man would be more likely, upon studying the facts, to conclude that choosing to believe is the appropriate choice, as opposed to choosing not to believe.

That was a concept I was really not prepared to hear. Until then, I don't think anyone had ever suggested to me that faith was a conclusion that one could arrive at on the basis of rational thought. I, and I suspect, many other scientists who've never really looked at the evidence, had kind of assumed that faith was something that you arrived at, either because it was drummed into your head when you were a little kid, or by some emotional experience, or some sort of cultural pressure. The idea that you would arrive at faith because it made sense, because it was rational, because it was the most appropriate choice when presented with the data, that was a new concept. And yet, reading through the pages of Lewis's book, I came to that conclusion over the course of several very painful weeks.

I didn't want this conclusion. I was very happy with the idea that God didn't exist, and had no interest in me. And yet at the same time, I could not turn away. I had to keep turning those pages. I had to keep trying to understand

---

[9] *Mere Christianity* is a theological book by C. S. Lewis, adapted from a series of BBC radio talks made between 1941 and 1944, while Lewis was at Oxford during World War II. Considered a classic of Christian apologetics, the transcripts of the broadcasts originally appeared in print as three separate pamphlets: *The Case for Christianity* (1942), *Christian Behaviour* (1942), and *Beyond Personality* (1944).

this. I had to see where it led. But I still didn't want to make that decision to believe.

The decision was an important step that I hadn't been aware of. You can argue yourself, on the basis of pure intellect, right up to the precipice of belief, but then you have to decide. I don't believe intellectual argument alone will push someone across the gap. Because we are not talking about something which can be measured in the same way the science measures the natural world, and then you decide what is natural truth. This is a supernatural truth. And in that regard, the spirit enters into this, not just the mind.

I struggled with that for many months, really resisting this decision, going forward, going backward. Finally, after about a year, I was on a trip to the northwest, and on a beautiful afternoon hiking in the Cascade Mountains, where the remarkable beauty of the creation around me was so overwhelming, I felt, "I cannot resist this another moment. This is something I have really longed for all my life without realizing it, and now I've got the chance to say yes." So I said yes. I was 27. I've never turned back. That was the most significant moment in my life.

To my surprise, I found myself fairly easily compelled by his (Lewis's) arguments about the existence of some sort of a God, because even as a scientist, I had to admit that we had no idea how the universe got started. The hard part for me was the idea of a personal God, who has and interest in humankind. And the argument that Lewis made there – the one that I think was most surprising, most earth-shattering and most life-changing is the argument about the existence of the moral law. How is it that we, and all other members of our species, unique in the animal kingdom, know what's right and what's wrong? In every culture one looks at, that knowledge is there.

Where did it come from? I reject the idea that that is an evolutionary consequence, because that moral law sometimes tells us that the right thing to do is very self-destructive. If I'm walking down the riverbank, and a man is drowning, even if I don't know how to swim very well, I feel this urge that the right thing to do is to try to save that person. Evolution would tell me exactly the opposite: preserve DNA. Who cares about the guy who's drowning? He's one of the weaker ones, let him go. It's your DNA that needs to survive. And yet that's not what's written within me.

Lewis argues that if you are looking for evidence of a God who cares about us as individuals, where could you more likely look than within your own heart at this very simple concept of what's right and what's wrong. And there it is. Not only does it tell you something about the fact that there is a spiritual nature that is somehow written within our hearts, but it also tells you something about the nature of God himself, which is that he is a good and holy God. What we have there is a glimpse of what he stands for[10].

I know this is not a new idea that Lewis came up with. It builds upon long traditions over centuries of careful scholarship and thought. But I'd never seen it before, and I don't think I've ever seen it explained as well as it is in his book.

---

[10] I would suggest that it is not only a moral law, but our capacity to love that shows us that there is a God. He has made us like Himself. Love is the basis of a moral law and the way we can understand His true nature.

> It (my decision to believe) certainly did carry with it this experience that life is now different. And along with that, this sense that God is not some distant concept, some ethereal, fuzzy entity. God became personal for me at that point. That really was the decision I was making, to believe not just in God, but in a God who wishes fellowship with me. That God is a God who both created the universe, and also had a plan that included me as an individual human being. And that he has made it possible for me, through this series of explorations, to realize that. It is not just a philosophy, it is a reality of a relationship."

So, if God is a God who loves us personally and is involved in our lives as He was in the Bible stories and in Francis Collins' life, then how can you listen to him? To begin our understanding of how God speaks to us today, we must start 6,500 years ago, at the beginning of the Bible.

Chapter 3. Ancient Interpreters

God speaks to us today through the ancient interpreters.

In the beginning there was no Bible, only God. But, as the Apostle John says:

> John 1:1-3 (NASB)
> 1 In the beginning was the Word, and the Word was with God, and the Word was God.
> 2 He was in the beginning with God.
> 3 All things came into being through Him, and apart from Him nothing came into being that has come into being.

So while there was no Bible, there was the Word, who is Jesus, the Son of God. One might say that Jesus is the physical aspect of God. According to John in the quote above, it was God working through Jesus who created everything including man and woman according to God's will, and he gave them the word. Those words became stories that were passed down by word of mouth throughout the generations until they were written down in the book of Genesis. Are these words historical facts and laws or are they something else?

If we look at Genesis Chapters 1 & 2, the beginning of the Bible, we find the precedent for the entire Bible.

There are contradictions.

As you can see below, Genesis Chapter 1 describes the sequence of creation as vegetation, the sun and moon, living creatures, and finally man and woman. Genesis Chapter 2 describes the sequence of creation as man, vegetation, living creatures, then and woman. See for yourself in the following excerpts. The highlighting is added for emphasis.

Genesis 1:9-31 (NASB)
9 Then God said, "Let the waters below the heavens be
gathered into one place, and let the dry land appear"; and
it was so.
10 God called the dry land earth, and the gathering of the
waters He called seas; and God saw that it was good.
11 Then God said, **"Let the earth sprout vegetation,**
plants yielding seed, and fruit trees on the earth
bearing fruit after their kind with seed in them"; and it
was so.
12 The earth brought forth vegetation, plants yielding
seed after their kind, and trees bearing fruit with seed in
them, after their kind; and God saw that it was good.
13 **There was evening and there was morning, a third**
day.
14 Then God said, "Let there be **lights** in the expanse of
the heavens to separate the day from the night, and let
them be for signs and for seasons and for days and years;
15 and let them be for lights in the expanse of the
heavens to give light on the earth"; and it was so.
16 God made the two great lights, the greater light to
govern the day, and the lesser light to govern the night;
He made the stars also.
17 God placed them in the expanse of the heavens to
give light on the earth,
18 and to govern the day and the night, and to separate
the light from the darkness; and God saw that it was good.
19 There was evening and there was morning, a **fourth**
day.
20 **Then** God said, "Let the **waters** teem with swarms of
living creatures, and let birds fly above the earth in the
open expanse of the heavens."
21 God created the great sea monsters and every living
creature that moves, with which the waters swarmed after
their kind, and every winged bird after its kind; and God
saw that it was good.

22 God blessed them, saying, "Be fruitful and multiply,
and fill the waters in the seas, and let birds multiply on the
earth."
23 There was evening and there was morning, a **fifth
day.**
24 **Then** God said, "Let the earth bring forth **living
creatures** after their kind: cattle and creeping things and
beasts of the earth after their kind"; and it was so.
25 God made the beasts of the earth after their kind, and
the cattle after their kind, and everything that creeps on
the ground after its kind; and God saw that it was good.
26 **Then** God said, **"Let Us make man in Our** (God and
Jesus)**, image according to Our likeness**; and let them
rule over the fish of the sea and over the birds of the sky
and over the cattle and over all the earth, and over every
creeping thing that creeps on the earth."
27 God created man in His own image, in the image of
God He created him; **male and female He created
them**...And there was evening and there was morning,
the sixth day.

Genesis 2:4-9 (NASB)
4 **This is the account of the heavens and the earth
when they were created**, in the day that the LORD God
made earth and heaven.
**5 Now no shrub of the field was yet in the earth, and
no plant of the field had yet sprouted,** for the LORD
God had not sent rain upon the earth, and there was no
man to cultivate the ground.
6 But a mist used to rise from the earth and water the
whole surface of the ground.
7 **Then the LORD God formed man of dust from the
ground**, and breathed into his nostrils the breath of life;
and man became a living being.

8 The LORD **God planted a garden** toward the east, in
Eden; and there He placed the man whom He had
formed.
9 Out of the ground the LORD God caused to grow every
tree that is pleasing to the sight and good for food; the
tree of life also in the midst of the garden, and the tree of
the knowledge of good and evil.

Genesis 2:18-22 (NASB)
18 **Then** the LORD God said, "It is not good for the man
to be alone; I will make him a helper suitable for him."
19 Out of the ground the LORD **God formed every beast
of the field** and every bird of the sky, and brought them to
the man to see what he would call them; and whatever the
man called a living creature, that was its name.
20 The man gave names to all the cattle, and to the birds
of the sky, and to every beast of the field, but for Adam
there was not found a helper suitable for him.
21 So the LORD God caused a deep sleep to fall upon
the man, and he slept; then He took one of his ribs and
closed up the flesh at that place.
22 The LORD **God fashioned into a woman** the rib
which He had taken from the man, and brought her to the
man.

So what is the right order of creation? Are these two conflicting stories written this way for a reason other than to describe the creation of the earth? Why wasn't one of these stories dropped over the centuries if they were inconsistent?

This brings us to the ancient interpreters.

Before written text, the word of God was handed down by word of mouth from generation to generation to people who were physically separated by time and distance. The best way to transmit information that cannot be recorded in any way was to create a story, poem, or song that can be remembered and retold around the campfire or in gatherings of a community. Eventually these stories were written down, copied, and re-copied for hundreds, and in some cases thousands, of years. It was not until about 300 AD/CE that the Hebrew and Christian texts began to consolidate into something resembling today's Bible. We will discuss this later. But as you can see, there were thousands of years of oral tradition and text passed on in communities of faith. Over time, ancient scribes consolidated the stories and edited early texts. The need grew within early Jewish communities to interpret what the texts meant and how they connected to each other. Believers felt compelled to pass the sacred texts on as unaltered as possible, but they needed a way to reconcile the contradictions so that they could understand what God was telling them and make them applicable to their day. Thus ancient interpretations were passed on along with the stories. As new interpretations were added, they were passed on as well. In a big picture view, that is how the Jewish Torah (five books of Moses), Tanah (sacred Jewish cannon texts including the Torah), and Talmud (oral tradition, interpretations, and rabbinical discussions) came to be.

It is here that I am going to recommend that you read *How to Read the Bible: A Guide to Scripture, Then and Now* by Dr. James L. Kugel[11]. This book discusses each of the great stories in the OT and describes how they were interpreted by the ancient interpreters and by modern scholars. This scholarly, but easily read, book wrestles with the differences between what people have believed for centuries and what scholars have learned over the last hundred and fifty years.

One great insight from Kugel's book is that God seems to have spoken to all the ancient interpreters in a similar way:

---

[11] *How to Read The Bible A Guide to Scripture, Then and Now,* James L. Kugel, Free Press, 2007

> " It is a striking fact that all ancient interpreters seem to have shared very much the same set of expectations about the biblical text. No one ever sat down and formulated these assumptions for them – they were simply assumed, just like our present-day assumptions about how we are to understand texts uttered by poets and group-therapy patients. However, looking over the vast body of ancient interpretations of different parts of the Bible, we can gain a rather clear picture of what their authors were assuming about the biblical text-- and what emerges is that, despite the geographic and cultural distance separating some of these interpreters from others, they all seem to have assumed the same four basic things about how the Bible was to be read:
>
> 1. The Bible was a fundamentally cryptic text: that is, when it said A, often it might really mean B.
> 2. The Bible was a book of lessons directed to readers in their own day. It might seem to talk about the past, but it is not fundamentally history. It is instruction, telling us what to do:
> 3. The Bible contained no contradictions or mistakes. It is perfectly harmonious, despite its being an anthology;
> 4. They believed that the entire Bible is essentially divinely given text, a book in which God speaks directly or through His prophets."[12]

These assumptions have lead to the interpretations of the Bible that are the foundations of our faith today. When I first read this, I realized that without being aware of it, these are also the same assumptions that I have used when I read the Bible. Could this say something about how God speaks to each of us? It is interesting that these four assumptions work together to allow different interpretations while being divinely inspired. Could this be how the Bible and the Holy Spirit work together to guide us?

---

[12] IBID page 14

If we return to the contradiction between Genesis Chapters 1 and 2 that we described before, we could hypothetically interpret it in the following way using the assumptions of the ancient interpreters.

Since there appears to be a contradiction in chapters 1 and 2 of Genesis, assumption 3 (totally consistent) would be violated. If assumption 3 cannot be violated, then, under assumption 1, we could make an interpretation that Chapter 1 is describing the creation of the universe from God's point of view, where Chapter 2 is describing God's creation of the Garden of Eden, Adam, and Eve. This could be a long time after the initial creation of the earth and all that is in it. The creation of the Garden of Eden and Adam and Eve could be an intervention on God's part to begin his personal involvement with people. That would solve the problem of that contradiction. But there is another.

What about the role of women? Assumption 2 states that these texts are directed to us today. Some could believe that since the Bible says women are formed from men then they must not be men's equals. And since the texts are divinely given under assumption 4, then this could be, and has been, interpreted that it is God's will that women should serve men. But is this interpretation God's will or is it the will of the interpreter? This interpretation is not what Genesis Chapter 1 says. Nor is it what Paul says:

> Galatians 3:26-29 (NASB)
> 26 For you are all sons of God through faith in Christ Jesus.
> 27 For all of you who were baptized into Christ have clothed yourselves with Christ.
> 28 There is neither Jew nor Greek, there is neither slave nor free man, **there is neither male nor female**; for you are **all one in Christ Jesus.**

29 And if you belong to Christ, then you are Abraham's descendants, heirs according to promise.

So, is our interpretation of Gods will for the service of women wrong, or is there a contradiction between Genesis 1, 2, and Galatians? As you can see, if we apply all 4 of the assumptions to the entire Bible, there are an infinite number of possibilities. One begins to see that, under these assumptions, the interpreter has the power to speak as God to those who will listen. Interpretation gives earthly power to the interpreter. This leads to a separation between interpreters and followers giving rise to rabbis, priests, bishops, and kings.

Chapters 1&2 of Genesis exist today because the people of each generation felt that they were sacred. I believe that they are. If the Bible is divinely inspired, the conclusion must be that God has placed contradictions in the Bible to engage us to converse with it. God uses these texts to speak to each of us, but the Bible text is only part of the message from God. Let's explore this further by looking at what Jesus has to say about how He and God speak to us.

Mark 4:2-12 (NASB)
2 And He was teaching them many things in **parables**, and was saying to them in His teaching,
3 "Listen to this! Behold, the sower went out to sow;
4 as he was sowing, some seed fell beside the road, and the birds came and ate it up.
5 "Other seed fell on the rocky ground where it did not have much soil; and immediately it sprang up because it had no depth of soil.
6 "And after the sun had risen, it was scorched; and because it had no root, it withered away.
7 "Other seed fell among the thorns, and the thorns came up and choked it, and it yielded no crop.

8 "Other seeds fell into the good soil, and as they grew up and increased, they yielded a crop and produced thirty, sixty, and a hundredfold."
9 And He was saying, **"He who has ears to hear, let him hear."**
10 As soon as He was alone, **His followers**, along with the twelve, began asking Him about the parables.
11 And He was saying to them, **"To you has been given the mystery of the kingdom of God**, but those who are outside get **everything** in parables,
12 so that WHILE SEEING, THEY MAY SEE AND NOT PERCEIVE, AND WHILE HEARING, THEY MAY HEAR AND NOT UNDERSTAND, OTHERWISE THEY MIGHT RETURN AND BE FORGIVEN."

What conclusions can we draw from this about how God speaks to us? We begin by using the ancient interpreters assumptions:

1. Jesus tells a story that contains a lesson that He wants us to learn. In this case it is an oral story. For us it is a biblical text that occurs in Matthew, Mark and Luke. The story is directed to the people to whom Jesus is speaking, not the participants in the story. We understand that Jesus is talking to us as well. The ancient interpreters Assumption 2 (a message for our day) must be true. Assumption 4 must also be true. Jesus is telling it so it must be divinely inspired.
2. He speaks in parables. So, assumption 1 must be correct. Jesus tells us in his discussion with the disciples after his sermon that the story is cryptic. It has a higher-level meaning that He shares with them.
3. There are contradictions between the story texts in Matthew, Mark, and Luke. There must, therefore, be something going on beyond the story itself as we questioned above. Jesus answers this in verse 9. **"He who has ears to hear, let him hear."** What does he

mean by this? He explains what he means in verse 11, **"To you has been given the mystery of the kingdom of God,"** i.e. "To you, my followers, who I will share the meaning I have intended for **you**."

4. What Jesus is really saying is that the story he told the masses is a parable. That story, as recorded in the Bible, has a meaning beyond the common understanding of the words. Jesus tells us that he interprets the parable to those whom he chooses. He did this in person for His disciples and He does this today through the Holy Spirit. When He says "he who has ears", He is actually saying He who listens to the Holy Spirit.

If what Jesus tells us is true, then this is also true of other stories in the Bible. The Bible confirms this throughout. For example:

> Psalms 78:1-3 (NASB)
> 1 Listen, O my people, to my instruction; Incline your ears to the words of my mouth.
> 2 I will open my mouth in a parable; I will utter dark sayings of old,
> 3 Which we have heard and known, And our fathers have told us.
>
> Ezekiel 17:1-2 (NASB)
> 1 Now the word of the LORD came to me saying,
> 2 "Son of man, propound a riddle and speak a parable to the house of Israel,
>
> Hosea 12:10 (NASB)
> 10 I have also spoken to the prophets, And I gave numerous visions, And through the prophets I gave parables.

For hundreds of years people have debated what the Bible is, how literally it should be read, and how it should be interpreted.

The Ancient Interpreters operated under a set of assumptions as they searched for **the** meaning of the scripture. Their history of interpretation helped both them and us come closer to God, but they missed one critical thing. They did not know about the Holy Spirit. There is no single interpretation of Biblical text. It can only be truly interpreted with the help of the Holy Spirit. I propose that the entire Bible is a collection of divinely given stories that are actually parables requiring our study and our listening to the Holy Spirit to fully understand. In addition the same Biblical text is used by the Spirit to provide understanding to a particular individual. That understanding might even change over time as the individual grows and his/her situation changes.

If there are contradictions in the Bible and the Bible is divinely inspired, it must be God who has allowed the contradictions to be there. Why would He do this? This gets to a central issue of this book. God could have caused the Bible to be written like the 10 Commandments. Do this, don't do that, believe the following; one message from one God to His people. He has not done this. Instead, He has provided us with a book that invites a dialogue - a book that causes us to search for the meaning and the answers. That is where the Holy Spirit plays a critical role. The Spirit works in the lives and thoughts of those who seek God. What we need is a shift in the common understanding of the Bible - from a book of Holy text to a book that facilitates a Holy conversation with God.

Chapter 4. Scholars

How shall we understand the Bible and the parables that it contains? Biblical scholars have been working on this since the beginning. In the last 150 years a combination of discoveries, research, and technologies have come together to provide us with valuable insights that can help us begin to answer part of this question. God speaks to us today through scholars.

The Councils of Hippo and Carthage canonized the Bible, as we know it in 397 CE. I believe that canonization was an inspired act of God. If there is a living God who loves us and He wants us to have a book that He can use to communicate with us, then the canonized Bible is the book as written. The insight for me was realizing that in spite of all that has been discussed and written, the words in the Bible are what people have read for over 1600 years. It is the book that God has used for hundreds of generations. The Bible, then, must contain the words that God wants us to read, but how we read them can be enlightened by what we have learned from scholars.

Discoveries such as the "Dead Sea Scrolls" and "Nag Hammadi" shed light on early Christian writings. Research into the history and cultures of the times and places mentioned in the Bible help our understanding by putting the Biblical text into context. The variety in use of language by people of different ages and geographic locations helped to identify where, when, and perhaps by whom the Biblical text was written. Technologies to aid archeology and reading ancient texts were instrumental in aiding research into the development of biblical text and its transmission from the original writer to what we see in our Bibles today. This research combined with thousands of theories makes up biblical scholarship.

I cannot possibly describe biblical scholarship in this book. I will give you a feel of what it looks like from a very big picture level and suggest that you begin more in-depth studies by reading three books - James Kugels[13] book on the Old Testament that I described previously, Diarmaid MacCulloch's Book *Christianity The First Three Thousand Years*[14], and *The Writings of St. Paul,* Wayne A. Meeks and John T. Fitzgerald, 2007 W.W. Norton & Company, Inc.[15] These books will not only summarize the scholarly debates and insights as we know them today, but will also show you the complexity and uncertainty that result from seeking finite answers about the Bible.

As a layman reader of the Bible, I could not understand why there are so many translations of the texts and why there are so many disagreements about what the Bible is saying.

A big problem is the language used to write the biblical text. What we call the Old Testament (OT) was originally written for the most part in Hebrew. The New Testament (NT) was written mostly in Greek. Ancient Hebrew contained no vowels or punctuation marks. Both were understood within the culture of the day and in the way the sentence was constructed. Ancient Hebrew writings are like lengthy complex "text messages" without vowels or pauses. A person of the day understood the text, but think about what someone from another time and place would have to know to understand today's cell-phone text messages.

---

[13] IBID

[14] *Christianity The First Three Thousand Years,* Diarmaid MacCulloch, 2009, Viking Penguin

[15] *The Writings of St. Paul,* Wayne A. Meeks and John T. Fitzgerald, 2007, W.W. Norton & Company, Inc.

In addition, in many cases the author could use several words to express his thought as we do in English. Scholars wonder why the author used a particular word 5 times in a discussion and then use a different word to apparently mean the same thing? One would assume that there is a reason. If we go back to our discussion about Genesis Chapters 1 & 2, we find that in Chapter 1 the author used the word 'elohim' for God where in Chapter 2 the word 'YHWH' or the "Lord God" is used. Is this because they are speaking of God in one case and Jesus or a different deity in another, or are there two different authors using different words for the same God? This is in part what confronts biblical scholars trying to interpret the OT.

Ancient Greek is another story. While it does have vowels and punctuation marks, each word has numerous meanings, some as many as 60. While you might know the Greek words of the text, there may not be a definitive meaning for the words. This results in debates between scholars who are interpreting the text from different frames of reference.

Picture this. Our Lord Jesus is born in the countryside of Galilee. He selects 12 ordinary people from fishermen to a tax collector to be his disciples. He begins to teach and perform miracles. At about age 33, almost 2,000 years ago, he is crucified. His disciples proclaim that he has risen from the dead and begin to spread the good news under fear of arrest from the Jews and Romans. The people surrounding Jesus were not documenting his life. In fact, they expected that he would return within their lifetime, so no reason to try to write things down, just tell as many people as they could.

How did the NT get written? While there are a lot of disagreements about the details, at the highest level, scholars agree on fundamental facts that help us to understand what God is telling us in the Bible. These facts are not commonly discussed in churches, yet they provide important insights to Bible readers.

First, the Gospels are not the earliest Christian written text. The NT is not written in chronological order. In fact, it is organized, for the most part, by the length of each book. Shortly after Jesus' crucifixion, in about CE 33, Saul of Tarsus (Paul) goes out to arrest people who believe in Jesus Christ. Paul is called by the risen Christ to be His apostle as described in Gal 1: 11-24. Without going to meet the other apostles in Jerusalem, Paul proceeds to preach in cities in what is now Turkey. In about 51 CE, after he has been preaching for about 14 years, Paul writes a letter to the Thessalonians. This is the first written Christian text that we have today. He then writes 6 other letters, Galatians 54 CE, First Corinthians, 53 – 55 CE, Second Corinthians 54-56 CE, Romans, 57 CE, Philippians 62 or 56 CE, ending with Philemon in 56 or 62 CE.

Second, all the other books attributed to Paul were probably not written by him, but in his name. In those days it was considered paying homage to write in another person's name that you admired. The Pastoral Letters, including First and Second Timothy and Titus, were most certainly written in 95-125 CE, after Paul's death. The dates are derived from historical events mentioned in the letters. In the case of the Pastoral Letters there is guidance to churches. In Paul's lifetime, churches consisted of groups of people meeting in people's homes. There were no Christian Churches. We will talk more about this when we discuss Paul.

Third, the Gospel of Mark is the oldest of the first three Gospels. It was written during the Jewish revolt against Rome that resulted in the destruction of the temple in 70 CE that sent the Jews fleeing from Jerusalem. This is also the time when Christians came to believe that Christ was not about to return immediately so they had better document the stories for future generations. Matthew and Luke had Mark's text available when they wrote theirs in about 85 CE and 90 CE respectively. Matthew was focused on Jewish Christians. Luke also wrote Acts and was focused on establishing the Christian church. The different audiences probably affected how these authors told the stories of Jesus. The similarities between these Gospels result from Matthew and Luke having Mark's text and another text that scholars have labeled Q. The lost Q Gospel is assumed because there are other similarities between Matthew and Luke that are not present in Mark. Mark is also assumed to be the first Gospel because it has the simplest wording. Later texts contain more detail and are therefore lengthier. This is explained in part by the idea that Matthew and Luke knew of oral stories that were not included in Mark. They came from a different frame of reference, and they were writing to different audiences, so they added the details that they felt were missing in Mark. If you want to understand the meaning of a story in Matthew, Mark, or Luke, look to Mark for the basic storyline.

Fourth, John's Gospel is the only one written by an apostle and eyewitness to Jesus' life. Scholars suggest 90 CE or later but perhaps about the same time as he wrote Revelation 54-68 CE, during Nero's reign. I should note that many scholars believe that two different Johns wrote the Gospel of John and Revelation. They base this on the fact that the writing style of the two books is quite different. However, by this time, John would have been quite old and would probably have used scribes to write at his direction. Obviously, there could have been two different scribes and the two books would have been written at different times. The Gospel was probably written in Ephesus, and Revelation was written on the Island of Patmos about 70 miles away. I believe that this explains the differences, but no one can be sure. John's writings did not get wide circulation until after his death so that may also explain the later date scholars' attribute to John. But, who better to relate the end-time revelation story than the Apostle that Jesus loved.

It is possible that Roman Emperor Nero played a more important role in creating John's writings and the Gospels then we acknowledge. He is known for his brutality against the Christians. He executed Peter and Paul. If we think about this for a moment, we can see how the early Christians must have felt. They believed that Christ would return during their lifetime. When Christ did not return to save Peter, Paul, and hundreds of Christians sacrificed in the Roman Circus, it would have been clear that He was not going to return soon. I believe that Nero's successful brutality against the Christians was the imperative for Jesus' followers to document Jesus' life, sayings, and resurrection. John was not rounded up and killed with the other disciples. Perhaps this is because he had escaped to Ephesus and perhaps later to Patmos. If this was the case, then Revelation may have been given to him not only to comfort the seven churches, but also to comfort John. By sending John, the churches, and us the follow-on to the resurrection story, Revelation, they and we could know that Christ will return as he said and reassure his followers.

John's Gospel is significantly different from the other Gospels. It is more spiritual and is aimed at telling believers what they need to know, not all that happened, as stated in John 20:31. After Jesus' death, Jesus brother, James, became the head of the Christian church in Jerusalem. This church was focused on telling Jews that their messiah had come. As a result, they had a hard time separating from the Jewish laws including circumcision. John's followers became separated from the Jerusalem church and those whom it taught. A paragraph from *Christianity The First Three Thousand Years* [16]sums this up.

> One might have expected that the result of this *(debate within the church as to what Jewish laws should be obeyed by Christians)* would be the development of two

---

[16] IBID Parenthesis added.

> branches of Christianity in fundamental disagreement with one another about their relationship with the parent Judaism: there would be a Jewish Church looking to the tradition represented by James *(Jesus brother)* and a Gentile Church treasuring the writings of Paul and John. In fact this is not so. There is one epistle in the New Testament which has been given James's name and which does represent a rather different view of the Christian life and the role of the Law from that of Paul, but otherwise all Christians alive today are the heirs of the Church which Paul created. The other type of Christianity once headed by the brother of the Lord has disappeared.

I believe that this is the work of the Holy Spirit who has done God's work to craft the Bible as we now know it. He is greater than we imagine.

Fifth, Bible texts have been edited and added to over time. In the past, my thought as to how the Bible was written matched the description in the first chapter of Revelation. I thought that God called the writer and inspired him to write what I was now reading except for the language translation. As it turns out, Revelation is the only book of the Bible that was written that way. In studying what the scholars have revealed, we now know that the Bible has a much more complex history than most people imagine. The OT was written from people's recollection of ancient stories. Eventually, these stories were written down, passed on, in some cases for thousands of years, and edited along the way. Of all the Hebrew Scriptures available to Jesus and his disciples in their lifetime, Isaiah is the most referenced by them. Here is an example from what Kugel has to say about the book of Isaiah

> "After all, if the original Isaiah's words were preserved, these scholars argue, they were probably first saved by the prophet's own inner circle—disciples of some sort,

perhaps even fellow prophets. Their decisions about how to edit and arrange his pronouncement ought not simply be dismissed. Even if they added to their master's words here and there or preserved them in some order other than the strictly chronological, perhaps this was because they were seeking to transmit a particular message that they felt the prophet had intended for the coming generations. As for still later material, some of it may have originated from people who were actually prophets themselves, recognized as such by their contemporaries, but whose words came to be incorporated, for one reason or another, in a book named for the prophet Isaiah." [17]:

If Isaiah has been added to and edited over the centuries prior to Jesus day, Jesus certainly thought that was acceptable because He quoted Isaiah in the Gospels. Isaiah's prophecies also took on new "Christian" meaning. Jesus used Isaiah to show people of his day that the prophecies of Isaiah were being fulfilled in their time.

Mark 1:1-3 (NASB)
1 The beginning of the gospel of Jesus Christ, the Son of God.
2 As it is written in Isaiah the prophet: "BEHOLD, I SEND MY MESSENGER AHEAD OF YOU, WHO WILL PREPARE YOUR WAY;
3 THE VOICE OF ONE CRYING IN THE WILDERNESS, 'MAKE READY THE WAY OF THE LORD, MAKE HIS PATHS STRAIGHT.'"

Mark 7:6-8 (NASB)
6 And He said to them, "Rightly did Isaiah prophesy of you hypocrites, as it is written: 'THIS PEOPLE HONORS

---

[17] *How to Read The Bible A Guide to Scripture, Then and Now,* James L. Kugel, Free Press, 2007, Page 564

ME WITH THEIR LIPS, BUT THEIR HEART IS FAR AWAY FROM ME.
7 'BUT IN VAIN DO THEY WORSHIP ME, TEACHING AS DOCTRINES THE PRECEPTS OF MEN.'
8 "Neglecting the commandment of God, you hold to the tradition of men."

John 12:36-40 (NASB)
36 "While you have the Light, believe in the Light, so that you may become sons of Light." These things Jesus spoke, and He went away and hid Himself from them.
37 But though He had performed so many signs before them, yet they were not believing in Him.
38 This was to fulfill the word of Isaiah the prophet which he spoke: "LORD, WHO HAS BELIEVED OUR REPORT? AND TO WHOM HAS THE ARM OF THE LORD BEEN REVEALED?"
39 For this reason they could not believe, for Isaiah said again,
40 "HE HAS BLINDED THEIR EYES AND HE HARDENED THEIR HEART, SO THAT THEY WOULD NOT SEE WITH THEIR EYES AND PERCEIVE WITH THEIR HEART, AND BE CONVERTED AND I HEAL THEM."

Now look at Isaiah and find these quotes that appear in capitols above. You will find that even Jesus adjusted the historical text to fit the meaning of his day. This shows that Jesus assumptions about the existing scripture were similar to those of the ancient interpreters.

Perhaps Isaiah said it best:

Isaiah 55:8-9 (NASB)
8 "For My thoughts are not your thoughts, Nor are your ways My ways," declares the LORD.

> 9 "For as the heavens are higher than the earth, So are My ways higher than your ways And My thoughts than your thoughts.

As for the NT, Jesus' disciples and the apostle Paul all thought that Christ would return in their lifetime so their focus was not initially on writing down what had happened. NT texts are letters from Paul, John, and early Christian disciples from the next generation. These texts have been preserved as sacred, yet edited and expanded to address early church, social, and political issues up to the point of their being canonized. Here is what Dr. MacCulloch has to say about Paul's books[18]:

> " Paul's admirers evidently decided to place increasing emphasis on his hierarchical view of Christian relationships and on his awareness of the scrutiny of Christian communities by non-Christians. Perhaps this was not surprising, as hopes of Christ's imminent return began to fade in the later first century and Christians began to realize that they must create structures that might have to last for a generation or more amid a world of non-believers. The change is visible in a series of further epistles which, although they assume the name of Paul, display a distinctive vocabulary and mechanically intensive reuse of phrases from his writings. They should be thought of as commentaries on or tributes to his impact and teaching."

These epistles / books are Ephesians, Colossians, 2 Thessalonians, 1&2 Timothy, and Titus.

Beyond changes to the text and the creation of books in the name of Paul and others, the Bible has been interpreted ever

---

[18] *Christianity The First Three Thousand Years*, Diarmaid MacCulloch, 2009, Viking Penguin, Page 118

since it was canonized by everyone who has studied it. You, me, your priest or minister, we are all getting new messages from the word of God. God seems to want it this way.

By the 13th Century there emerged the notion of a fourfold interpretation where every verse might have four different meanings simultaneously; literal, allegorical, moral, and anagogical (future / end-time). Here is Kugel's description:

> The literal sense teaches the facts (or "deeds"); the allegorical, what you should believe; the moral sense, what you should do; and the anagogical, where you are headed. The standard example of the application of this fourfold reading of Scripture is that of the city of Jerusalem as it might appear in a biblical verse. According to the literal (or historical) sense, Jerusalem refers to an actual city, a place where the Jews dwelt in biblical times. According to the allegorical sense, however Jerusalem refers to the church, so that when a biblical verse talks about "dwelling in Jerusalem," it might really mean "abiding in the Church." The moral sense (sometimes also called the topological sense) focuses more on the life of the individual soul- so that, in the moral reading, Jerusalem itself might be taken to represent a person's soul and therefore teach, "what you (as an individual) should do." The anagogical, or eschatological, sense teaches about what is to be in the end-time – hence, Jerusalem here might represent the heavenly city of God that will be revealed in the fullness of time. [19]

In 1985 – 95, 150 biblical scholars began an effort to find the text that was unaffected by revision. They called it the "Search for the Historical Jesus". The result of this effort resembled pealing back

---

[19] *"How to Read The Bible A Guide to Scripture, Then and Now",* James L. Kugel, Free Press, 2007 Page 23

the layers of an onion. When you look critically at all of the scripture that has been attributed to Jesus, you find that, there is no there, there. All that we know for sure is, as Paul says, Jesus, Him crucified and raised from the dead. Yet, Jesus and His words in the Bible have impacted billions of people's lives since He lived and was crucified and rose from the dead. Proof enough of God's power in people's lives and of the value of the Bible.

Scholars are like scientists. They study what they can see, hear, smell, taste, and feel. What they have to work with is the text that has been canonized, other ancient texts, the language and words used, how the books have been put together, the process of transmission from the time they were written to today, the cultures and world events that were transpiring during this period, and other historical writings from the Greeks, Romans and others. Their conclusions are based on what the physical evidence and scientific technologies can tell us. These are important insights, but only part of how God communicates with us.

Modern scholars are beginning to see that studying biblical text and looking into its past is only part of the story. Dr. Walter Brueggemann's presentation at the 41st Trinity Institute National Theological Conference describes what scholars are beginning to see.

> "One helpful way into such Jewish modes that refuse closure is to pay attention to Freud's theory and practice of psychoanalysis, for it is clear enough that Freud's way of "reading" repressed personhood was taken from rabbinic ways of reading the hiddenness of texts. As you know, the psychotherapeutic conversation can walk endlessly around a memory, an event, a dream, a phrase, because the memory, event, dream or phrase has a rich capacity for multiple meanings, any of which may be censored, but all of which may be emancipatory.

Freud's great insight is that the human self, never more than partially brought to awareness, is thick, layered, and conflicted:

-The self is thick in the sense that word and image and memory are freighted with more meaning and force than any single saying of it can ever unpack.

-The self is layered because, in our fearful repression, we have stacked experience upon experience, hurt upon hurt, rage upon rage, and sometimes even joy upon joy.

-The self is conflicted because of the inescapable friction between the felt self and the socially expected self, a friction that we spend our lives negotiating. And of course what we present (even to ourselves) of such a complex self is most often only in bits and pieces, only some of which come to our awareness.

If, as I believe following Susan Handleman[20], Freud learned this approach from rabbinic ways of reading texts to watch for their thickness, layered quality, and conflictedness, we may take his great insight about the modern self back into our reading of biblical texts that are also thick, layered, and conflicted.

At its best, criticism has seen exactly that about the text:

-The text is thick; that is what keeps preachers going. The text, in rich artistry, cannot be flatly explained or reduced to single meaning, even though pre-critical innocence and critical rationality tried to do so. You can check this by

---

[20] Dr. Susan Handelman is a professor of English at Bar-Ilan University in Israel. A native of Chicago, she is the author of two books on Jewish thought and a translator of the Rebbe's book "On the Essence of Chassidus."

reading the old historical critical commentaries. What strikes one about them is how thin they are; when the text is taken thinly, there is almost nothing to be said about texts.

-The text is layered most broadly in the JEDP hypothesis[21], so that the text, like the self, is an assemblage of many voices over time, each of which has struggled to be the last word.

-The text is conflicted, so that it is not difficult to discern tensions and contradictions in the text. When honored, those tensions and contradictions are not to be explained away; rather they are the matrix of generative interpretation as with the work of pastoral therapy.

A pre-critical certainty wants to deny this quality of the text. A critical understanding wants to explain and resolve and sort out. But in good psychotherapy, after the work of "explanation" has been attempted, there comes a time of pause for wonderment, silence, and respect for the rich dimensions of the self in front of us, "the many selves of the self." So it is with the text. There comes a time for a pause in wonderment, silence, and respect for a text that refuses thinness. We dare to say, moreover, that it is precisely this text, in its thickness, layeredness, and conflictedness, that bears witness to the true God, this Jewish God who refuses the old Greek rationality and who continues to refuse our modernist thinness, this triune God who refuses the old

---

[21] The documentary or JEPD hypothesis assumes that the text of the Torah as preserved can be divided into identifiable sources that predate its compilations by centuries, the Jahwist (J) source being the oldest, dating to as early as the 10th century BCE, along with the Elohist (E), the Deuteronomist (D), and the Priestly source (P), dating to the 8th to 6th centuries. The final compilation of the extant text is dated to either the 6th or 5th century BCE. (Wickipedia)

> deistic flatness of monarchy. We dare say, beyond that as we move from the text to the God witnessed there, that we are in the image of that God, thick, layered, and conflicted. So we can imagine that any congregation, meeting around this thick, layered, conflicted text, bears witness to this God in whom we live and move and have our being. We do so in resistance against the reductionism of pre-critical innocence, in refusal of Enlightenment rationality, and in defiance of the thin requirements of technological consumerism. The entire interpretive project is open and deep and emancipatory."[22]

I began this book by saying that God is greater than we imagine. The scholars have proven that the simplistic view of the books of the Bible being written by a single inspired person is not true. The truth is far greater than these simplistic notions. God has worked with all the Biblical writers, editors, interpreters and publishers to create a living document that the Holy Spirit can use to speak to each one of us. Scholars have shown us is that God has worked through many, many people over hundreds, and even thousands of years to provide us with the sacred book we call the Bible. But the story doesn't end here.

---

[22] *Trinity Institute® January 19-21, 2011 Walter Brueggemann: Where is the Scribe?, Page 9 (http://www.trinitywallstreet.org/faith/institute/2011/transcripts)*

Chapter 5. Paul

From the scholars, we now know that Paul created the original Christian texts. God speaks to us today through the letters that Paul wrote.

Scholars say that the first letter we have from Paul is 1 Thessalonians. It was written about 51CE and was probably written about 14 years or so after his revelation and conversion. Philemon, perhaps the last letter we have from him, was written around the year 55-62CE. In it, Paul describes himself as an old man. Paul probably died in 64CE during Nero's reign. This makes Paul a contemporary of Jesus and His original disciples. Further, scholars say that the gospel that Paul preached has become Christ's church on earth. But has it? Now that we know that he actually only wrote 7 letters, let's look at those to see what he was really saying. This may actually require us to unlearn some of what we have been taught from reading all the letters attributed to Paul.

While Paul was a contemporary of Jesus, he did not write about Jesus the man. His focus stemmed from his calling by Christ after He rose from the dead. Let's begin by seeing what Paul's gospel really is by looking at what we know of his calling from Christ.

Paul writes in Galatians and 2 Corinthians about his calling from Christ:

> Galatians 1:11-24 (NASB)
> 11 For I would have you know, brethren, that the gospel which was preached by me is not according to man.
> 12 For I neither received it from man, nor was I taught it, but I received it through a revelation of Jesus Christ.

13 For you have heard of my former manner of life in Judaism, how I used to persecute the church of God beyond measure and tried to destroy it;
14 and I was advancing in Judaism beyond many of my contemporaries among my countrymen, being more extremely zealous for my ancestral traditions.
15 But when God, who had set me apart even from my mother's womb and called me through His grace, was pleased
16 to reveal His Son **in me** so that I might preach Him among the Gentiles, **I did not immediately consult with flesh and blood,**
17 nor did I go up to Jerusalem to those who were apostles before me; **but I went away to Arabia,** and returned once more to Damascus.
18 Then three years later I went up to Jerusalem to become acquainted with Cephas, and stayed with him fifteen days.
19 But I did not see any other of the apostles except James, the Lord's brother.
20 (Now in what I am writing to you, I assure you before God that I am not lying.)
21 Then I went into the regions of Syria and Cilicia.
22 I was still unknown by sight to the churches of Judea which were in Christ;
23 but only, they kept hearing, "He who once persecuted us is now preaching the faith which he once tried to destroy."
24 And they were glorifying God because of me.

And, later on, Paul tells us:

2 Corinthians 12:1-6 (NASB)
1 Boasting is necessary, though it is not profitable; but I will go on to visions and revelations of the Lord.

2 I know a man in Christ who fourteen years ago—whether in the body I do not know, or out of the body I do not know, God knows—such a man was caught up to the third heaven.
3 And I know how such a man—**whether in the body or apart from the body I do not know, God knows—**
4 was caught up into Paradise and heard inexpressible words, which a man is not permitted to speak.
5 On behalf of such a man I will boast; but on my own behalf I will not boast, except in regard to my weaknesses.
6 For if I do wish to boast I will not be foolish, for I will be speaking the truth; but I refrain from this, so that no one will credit me with more than he sees in me or hears from me.

While this is Paul's own story, most churches focus on Luke's account of Paul's revelation in Acts, not Paul's own words. Now let's see what Acts / Luke has to say:

Acts 9:1-28 (NASB)
1 Now Saul, still breathing threats and murder against the disciples of the Lord, went to the high priest,
2 and asked for letters from him to the synagogues at Damascus, so that if he found any belonging to the Way, both men and women, he might bring them bound to Jerusalem.
3 As he was traveling, it happened that he was approaching Damascus, and **suddenly a light from heaven flashed around him;**
4 and he fell to the ground and heard a voice saying to him, "Saul, Saul, why are you persecuting Me?"
5 And he said, "Who are You, Lord?" And He said, "I am Jesus whom you are persecuting,

6 but get up and enter the city, and it will be told you what you must do."

7 The men who traveled with him stood speechless, hearing the voice but seeing no one.

8 Saul got up from the ground, and though his eyes were open, he could see nothing; and leading him by the hand, they brought him into Damascus.

9 And he was three days without sight, and neither ate nor drank.

10 Now there was a disciple at Damascus named Ananias; and the Lord said to him in a vision, "Ananias." And he said, "Here I am, Lord."

11 And the Lord said to him, "Get up and go to the street called Straight, and inquire at the house of Judas for a man from Tarsus named Saul, for he is praying,

12 and he has seen in a vision a man named Ananias come in and lay his hands on him, so that he might regain his sight."

13 But Ananias answered, "Lord, I have heard from many about this man, how much harm he did to Your saints at Jerusalem;

14 and here he has authority from the chief priests to bind all who call on Your name."

15 But the Lord said to him, "Go, for he is a chosen instrument of Mine, to bear My name before the Gentiles and kings and the sons of Israel;

16 for I will show him how much he must suffer for My name's sake."

17 So Ananias departed and entered the house, and after laying his hands on him said, "Brother Saul, the Lord Jesus, who appeared to you on the road by which you were coming, has sent me so that you may regain your sight and be filled with the Holy Spirit."

18 **And immediately there fell from his eyes something like scales, and he regained his sight, and he got up and was baptized;**

19 and he took food and was strengthened. Now for several days he was with the disciples who were at Damascus,
20 and immediately he began to proclaim Jesus in the synagogues, saying, "He is the Son of God."
21 All those hearing him continued to be amazed, and were saying, "Is this not he who in Jerusalem destroyed those who called on this name, and who had come here for the purpose of bringing them bound before the chief priests?"
22 But Saul kept increasing in strength and confounding the Jews who lived at Damascus by proving that this Jesus is the Christ.
23 When many days had elapsed, the Jews plotted together to do away with him,
24 but their plot became known to Saul. They were also watching the gates day and night so that they might put him to death;
25 but his disciples took him by night and let him down through an opening in the wall, lowering him in a large basket.
26 When he came to Jerusalem, he was trying to associate with the disciples; but they were all afraid of him, not believing that he was a disciple.
27 But Barnabas took hold of him and brought him to the apostles and described to them how he had seen the Lord on the road, and that He had talked to him, and how at Damascus he had spoken out boldly in the name of Jesus.
28 And he was with them, moving about freely in Jerusalem, speaking out boldly in the name of the Lord.

In Paul's own letter, he tells us that he was caught up to heaven and heard inexpressible words. He then went to Arabia without speaking to the other apostles or going to Jerusalem. Luke reports that he came to Jerusalem from Damascus where he had his revelation from Christ, and had difficulty meeting with the apostles until Barnabas helped him. Could it be that Luke wanted to project a closer relationship between Paul and the original apostles than really existed?

In the footnotes to Meeks book[23], he points out that modern romantic notions of Paul's meditating in the desert are off the mark; more likely he was preaching in the Hellenized cities like Petra and Bostra- otherwise the hostility of the Nabatean ethnarch, who pursued him back to Damascus (2 Cor 11:32), is inexplicable. The Acts' account of this period are incomplete, resulting in a distorted picture of Paul's relationship to Jerusalem. So whom do we believe- Paul in his own letters to the churches or Luke who wrote his Gospel and Acts sometime after 70CE but before 150CE? The answer has to be Paul.

If we take what we have learned from the scholars that Paul only wrote 7 letters and that his is the gospel that we follow because God has let it prosper over others, then we need to see what he is telling us and recognize that the other books written in his name are focused on the establishment of the Christian church by followers later in the story. Establishment of the Church was critical to sheltering God's people during persecutions in its early days, bringing God's message to the world, and preserving the sacred text until today. The letters written in Paul's name are undoubtedly God-inspired for these purposes.

Paul had a direct eyewitness revelation from Christ. His gospel message from Christ is in his original letters. So what is Paul's gospel? What was he telling the early churches and us today? Read his letters yourself for the full story. Here are some excerpts focusing on the story of how God speaks to us. I am

---

[23] *The Writtings of St. Paul,* Wayne A. Meeks & John T. Fitzgerald, W.W.Norton & Company New York 2007, 1972, Page 13.

presenting them in the order that the scholars believe that he wrote the letters (highlights added for emphasis):

1 Thessalonians 1:4-5 (NASB)
4 knowing, brethren beloved by God, His choice of you;
5 for our gospel did not come to you in word only, but also in power and in the **Holy Spirit** and with full conviction; just as you know what kind of men we proved to be among you for your sake.

Galatians 3:24-29 (NASB)
24 Therefore the (Jewish) Law has become our tutor to lead us to Christ, so that we may be justified by faith.
25 But now that faith has come, we are no longer under a tutor.
26 For **you are all sons of God** through faith in Christ Jesus.
27 For all of you who were baptized into Christ have clothed yourselves with Christ.
28 There is neither Jew nor Greek, there is neither slave nor free man, **there is neither male nor female; for you are all one in Christ Jesus.**
29 And if you belong to Christ, then you are Abraham's descendants, heirs according to promise.

1 Corinthians 1:22-25 (NASB)
22 For indeed Jews ask for signs and Greeks search for wisdom;
23 but **we preach Christ crucified**, to Jews a stumbling block and to Gentiles foolishness,
24 but to those who are the called, both Jews and Greeks, Christ the power of God and the wisdom of God.
25 Because the foolishness of God is wiser than men, and the weakness of God is stronger than men.

1 Corinthians 2:10-16 (NASB)
10 For to us God **revealed them through the Spirit**; for the Spirit searches all things, even the depths of God.
11 For who among men knows the thoughts of a man except the spirit of the man which is in him? Even so the thoughts of God no one knows except the Spirit of God.
12 Now we have received, not the spirit of the world, but the Spirit who is from God, so that we may know the things freely given to us by God,
13 which things we also speak, not in words taught by human wisdom, but in those **taught by the Spirit, combining spiritual thoughts with spiritual words.**
14 But a natural man does not accept the things of the Spirit of God, for they are foolishness to him; and he cannot understand them, because they are spiritually appraised.
15 **But he who is spiritual appraises all things**, yet he himself is appraised by no one.
16 For WHO HAS KNOWN THE MIND OF THE LORD, THAT HE WILL INSTRUCT HIM? But we have the mind of Christ.

1 Corinthians 11:23-26 (NASB)
23 For **I received from the Lord** that which I also delivered to you, that the Lord Jesus in the night in which He was betrayed took bread;
24 and when He had given thanks, He broke it and said, "This is My body, which is for you; do this in remembrance of Me."
25 In the same way He took the cup also after supper, saying, "This cup is the new covenant in My blood; do this, as often as you drink it, in remembrance of Me."
26 For as often as you eat this bread and drink the cup, you proclaim the Lord's death until He comes.

1 Corinthians 12:7-11 (NASB)
7 But to **each one is given the manifestation of the Spirit** for the common good.
8 For to one is given the word of wisdom through the Spirit, and to another the word of knowledge according to the same Spirit;
9 to another faith by the same Spirit, and to another gifts of healing by the one Spirit,
10 and to another the effecting of miracles, and to another prophecy, and to another the distinguishing of spirits, to another various kinds of tongues, and to another the interpretation of tongues.
11 But one and the same Spirit works all these things, distributing to **each one individually** just as He wills.

1 Corinthians 15:3-11 (NASB) - Paul's gospel
3 For I delivered to you as of **first importance** what I also received, that Christ died for our sins according to the Scriptures,
4 and that He was buried, and that He was raised on the third day according to the Scriptures,
5 and that He appeared to Cephas, then to the twelve.
6 After that He appeared to more than five hundred brethren at one time, most of whom remain until now, but some have fallen asleep;
7 then He appeared to James, then to all the apostles;
8 and last of all, as to one untimely born, He appeared to me also.
9 For I am the least of the apostles, and not fit to be called an apostle, because I persecuted the church of God.
10 But by the grace of God I am what I am, and His grace toward me did not prove vain; but I labored even more than all of them, yet not I, but the grace of God with me.
11 Whether then it was I or they, so we preach and so you believed.

2 Corinthians 3:2-3 (NASB)
2 You are our letter, written in our hearts, known and read by all men;
3 being manifested that you are a letter of Christ, cared for by us, written not with ink but with the **Spirit** of the living God, not on tablets of stone but on tablets of human hearts.

2 Corinthians 3:17-18 (NASB)
17 **Now the Lord is the Spirit, and where the Spirit of the Lord is, there is liberty.**
18 But we all, with unveiled face, beholding as in a mirror the glory of the Lord, are being transformed into the same image from glory to glory, just as from the Lord, the Spirit.

Romans 8:22-39 (NASB)
22 For we know that the whole creation groans and suffers the pains of childbirth together until now.
23 And not only this, but also **we ourselves, having the first fruits of the Spirit,** even we ourselves groan within ourselves, waiting eagerly for our adoption as sons, the redemption of our body.
24 For in hope we have been saved, but hope that is seen is not hope; for who hopes for what he already sees?
25 But if we hope for what we do not see, with **perseverance** we wait eagerly for it.
26 In the same way the Spirit also helps our weakness; for we do not know how to pray as we should, but the **Spirit Himself intercedes for us** with groanings too deep for words;
27 and He who searches the hearts knows what the mind of the Spirit is, because He intercedes for the saints according to the will of God.
28 And we know that God causes all things to work together for good to those who love God, to those who are called according to His purpose.

29 For those whom He foreknew, He also predestined to become conformed to the image of His Son, **so that He would be the firstborn among many brethren;**
30 and these whom He predestined, He also called; and these whom He called, He also justified; and these whom He justified, He also glorified.
31 What then shall we say to these things? If God is for us, who is against us?
32 He who did not spare His own Son, but delivered Him over for us all, how will He not also with Him freely give us all things?
33 Who will bring a charge against God's elect? God is the one who justifies;
34 who is the one who condemns? Christ Jesus is He who died, yes, rather who was raised, who is at the right hand of God, who also intercedes for us.
35 Who will separate us from the love of Christ? Will tribulation, or distress, or persecution, or famine, or nakedness, or peril, or sword?
36 Just as it is written, "FOR YOUR SAKE WE ARE BEING PUT TO DEATH ALL DAY LONG; WE WERE CONSIDERED AS SHEEP TO BE SLAUGHTERED."
37 But in all these things we overwhelmingly conquer through Him who loved us.
38 For I am convinced that neither death, nor life, nor angels, nor principalities, nor things present, nor things to come, nor powers,
39 nor height, nor depth, nor any other created thing, will be able to separate us from the love of God, which is in Christ Jesus our Lord.

Romans 14:14 (NASB)
14 I know and am convinced in the Lord Jesus that nothing is unclean in itself; **but to him who thinks anything to be unclean, to him it is unclean**.

Romans 15:4 (NASB)
4 For whatever was written in earlier times was written for **our** instruction, so that through perseverance and the encouragement of the Scriptures we might have hope.

Philippians 2:13 (NASB)
13 for it is God who is at work **in you,** both to will and to work for His good pleasure.

Paul defines the basic tenets of Christianity as we know it today. Some things we seem to have overlooked.

1. 1 Corinthians 15:3-8 is Paul's gospel.
2. He gives us the Mass / Communion ritual in 1 Corinthians11:23-26 that he says he received directly from the Lord.
3. Romans 15:4 declares that OT texts were written for instruction today and are not necessarily histories or focused on the people who initially read them.
4. Galatians 3:28 says that we are all one in Christ, so men and women are equal according to Paul. The scriptures that are so male-oriented in Timothy and other pastoral scripture were not written by Paul. They were probably written to help the church fit in with the culture of the day yet claim the authority of Paul.
5. Romans 8:22-39 describes God's master plan for us. It tells us that our ultimate destiny is to become a son of God in Christ's image.
6. Romans 14:14 makes it clear that one must follow their beliefs or he / she is not honoring God.
7. Read through the text excerpts again with particular attention to the scripture that I highlighted. Notice Paul's emphasis on the work of the Spirit. He is telling us that the Spirit of God is in those who believe. This is largely missing in the writings attributed to him. Also notice that he tells us that the Spirit teaches us individually through a

> combination of "spiritual thoughts with spiritual words" (1 Corinthians 2:10-16). Spiritual thoughts are those that come to you from the Holy Spirit, whereas spiritual words are the inspired words in the Bible. Here Paul is telling us that God communicates with us through insights from the Spirit that should be tested or consistent with the Biblical text. It is possible to receive the spirit of the world (human thoughts that come from within you and others), but the way to know that the Holy Spirit is speaking to you is to see if the thoughts are consistent with the Bible.

The preceding are the fundamental teachings of Paul. Taken together, Paul's message that is based on his direct revelation from Christ revolutionizes our understanding of God's relationship with us. Paul is saying that if we believe his gospel, then the result will be that God will put his Spirit within you. By listening to the voice of the Spirit (spiritual thoughts) within the context of the canonized Bible (spiritual words), he will establish a personal relationship with you and you will be saved through Jesus Christ.

However, if your church, sect, or personal belief demands something in addition to Paul's gospel, you must follow that belief as well as Paul says in Romans 14:14,

> " I know and am convinced in the Lord Jesus that nothing is unclean in itself; **but to him who thinks anything to be unclean, to him it is unclean."**

This is a concept that most churches don't talk about. It is basically saying that God wants you to love him and obey him. If you believe that something is necessary to obey God and you don't do it, then you are not showing love to God even if He does not demand it. So, if you believe you need to be circumcised, or worship three times a day, or eat fish on Friday to be a good Christian, then you do. However, there is nothing in the gospel that requires this. Therefore, if you believe in the gospel, and work to grow your personal relationship with God day by day, then you are free of the requirement to be circumcised, etc.

Paul's message is consistent with Jesus words in the Gospels. The following are three statements in Jesus' words on the subject of the burdens that the Pharisees were putting on the Jews. If the Pharisees said something to the people it became a requirement that the people needed to obey if they were honoring their Jewish faith.

> Matthew 23:13 (NASB)
> 13 "But woe to you, scribes and Pharisees, hypocrites, because you shut off the kingdom of heaven from people; for you do not enter in yourselves, nor do you allow those who are entering to go in.
>
> Mark 7:6-8 (NASB)
> 6 And He said to them, "Rightly did Isaiah prophesy of you hypocrites, as it is written: 'THIS PEOPLE HONORS ME WITH THEIR LIPS, BUT THEIR HEART IS FAR AWAY FROM ME.
> 7 'BUT IN VAIN DO THEY WORSHIP ME, TEACHING AS DOCTRINES THE PRECEPTS OF MEN.'
> 8 "Neglecting the commandment of God, you hold to the tradition of men."
>
> Luke 11:46 (NASB)
> 46 But He said, "Woe to you lawyers as well! For you weigh men down with burdens hard to bear, while you yourselves will not even touch the burdens with one of your fingers.

So, Paul's teachings are contained in the letters he personally wrote. He wrote these letters beginning about 14 years after he started preaching. The last was written about 6 years before he died.

How did Paul's followers react and what became of his efforts? The answer is that Paul was interpreted in many different ways after his death, each claiming Paul as the founder of their faith. Some followers focused on Paul's emphasis on the Spirit's work within believers. They were called Gnostics. Other churches focused on the written texts as the source of their belief. They became the orthodox faith and have been labeled antignostics because they believed that a more literal reading of the scripture was essential. Here is what Dr. Elaine H. Pagels has to say in her essay "The Gnostic Paul"[24] concerning these two major sects that evolved from Paul's work:

Dr. Pagels says that after Paul's death, between the first and second centuries CE, two opposing traditions of Pauline exegesis / understanding emerged. Each claimed to be the authentic message from Christ and Paul yet one was antignostic while the other was labeled Gnostic.

The antignostic view was formed primarily from Paul's Pastoral Letters. This view focused on the biblical texts and became the orthodox teaching of the Church. It interpreted Paul as labeling the Gnostic, or spirit-filled people, as "false teachers" and branded them as heretics. The antignostic / orthodox view focused on Paul's role as an organizer of churches. This view is also supported by the admonitions Paul had to the "Strong" people of Corinth who claimed to have knowledge / wisdom from the Spirit that other members of the church who were not "Spirit filled" did

---

[24] Elaine H. Pagels, *The Gnostic Paul (Philadelphia: Fortress, 1975; rep.Philadelphia: Trinity Press Internaitonal 1992; rep The Writtings of St. Paul,* Wayne A. Meeks & John T. Fitzgerald, W.W.Norton & Company New York 2007, 1972) Pages 274-283.

not understand. It was essentially the universal view of Paul's teaching verses the elitist view.

The Gnostic view was based on the letters that scholars believe Paul actually wrote. This view is that Paul taught in two ways at once. On one hand he preached to the multitudes the gospel of Christ crucified and raised in a way that they could understand him, "according to the flesh". But to the "elect", he proclaimed Christ "according to the spirit" and that "each one knows the Lord in his own way: and not all know Him alike." The argument with the "orthodox" was that they read scripture too literally where it should also be read symbolically as Paul intended. They claimed that from the beginning Christ spoke in parables so that the many / the psychic / the called would hear, but not understand. Jesus would only explain the true meaning of the parables to the disciples who were the chosen / the pneumatic / the elect. The Gnostic view was that Paul's letters spoke in two ways at once: To the psychics- in terms they can grasp, the obvious message of the letters, to the elect- there was a deeper communication hidden in images of his letters. This deeper communication is the "living water" Christ offers to the elect who are the only ones capable of perceiving it and yearn to learn it.

These two views clashed as the Christian community wrestled with organizing the Church and canonize the Bible. Claims of exclusivity and secret knowledge beyond the scriptures politically doomed the Gnostic's chances of becoming the orthodox view, and resulted in the Christian community branding this form of belief as heretical. However, texts discovered at Nag Hammadi offers evidence for a gnostic Pauline tradition.

In actuality, both the second-century orthodox Christians and the Gnostics, like the Valentinians, have something to offer. Both have missed the subtlety that lies in-between.

Dr. Pagels makes several points on behalf of the Gnostics that are important to consider in understanding how God speaks to us today. The first point to mention is that the Gnostics believed that the Holy Spirit is how God speaks to us. The Spirit is part of our belief in the Trinity. John tells us that Jesus breathed on the Disciples and gave them the Holy Spirit. Acts describes the Holy Spirit being given to the Apostles. Paul speaks of the Spirit in his letters and as the source of his revelation, and yet, the working of the Spirit today is an issue that is seldom discussed in traditional churches. Perhaps that is because the work of the Spirit is within us and is not verifiable in terms that can be proved. It may also reflect a discomfort of Pentecostal churches by traditional churches that have become prominent in the early 20th century especially in Latin America.

The second point is that gnostic followers of Valentinians focused on the letters that scholars believe Paul actually wrote and dismissed the others. It is clear from those letters, that the Spirit played a major role in Paul's life and teaching. It seems that the Pastoral Letters were influenced by the arguments between the Gnostics and antignostics when they were written after Paul's death. It also seems clear that Paul's interaction with the Spirit directed his apostleship including where and when he went to preach.

> Romans 1:9-10 (NASB)
> 9 For God, whom I serve in my spirit in the preaching of the gospel of His Son, is my witness as to how unceasingly I make mention of you,
> 10 always in my prayers making request, if perhaps now at last by the will of God I may succeed in coming to you.

The third point is the idea that the Valentinians believed that the scriptures should not only be read literally, but symbolically as well. Earlier we described the four assumptions of the ancient interpreters. These assumptions are embodied in Valentinian belief. Three of these assumptions are that the Bible is fundamentally cryptic (i.e. symbolic in Valentinian terms), contain lessons for the reader's day (inspired by the spirit's work within the reader), and that the Bible is divinely inspired in which God speaks directly or through His prophets to readers of the Bible (and continues to do so both in what is written and what the Spirit inspires the reader to think as he or she searches for understanding). Paul certainly makes this point in Romans. The problem with this belief is that gnostic church leaders viewed Spiritual insights as universal not personal. This could have resulted in Christian churches following the spiritual insights of charismatic leaders not the scriptures as their base.

The fourth point is that the orthodox Christians seemed to dismiss the Spirit, while acknowledging its existence. The Valentinians seemed to focus on secrets that come from the Spirit and oral tradition, while describing what the scripture tells us as food for children. This resulted in their differentiation between the "Called" and the chosen "Elect". Both groups miss the point in between. Believers should not consider the scriptures secondary to spiritual inspiration and focus on ideas that come from people thinking that they have a secret truth. Both the Bible and the Spirit come from God. They will work together to bring Gods message to each of us.

These four points show that the Gnostics had some good points to offer in the debate over what belief should be considered orthodoxy in the early days of the church. However, many of the ancients who held gnostic assumptions would have made sure that they kept their distance from the Gnostics. By the fourth and fifth century, the emerging Catholic and Orthodox churches, rightly or wrongly, had declared Gnostic beliefs firmly beyond the pale of orthodoxy to the point of death.

Whether Paul should be labeled a Gnostic or not is looking backwards from a more modern definition of the term. The antignostic attack on the Gnostics of the first century was an important political struggle to pull the Church together, to keep it focused on the texts that were written, and not splinter into various groups claiming different spiritual insights. Oddly, I think that the Holy Spirit wanted it that way because the orthodox Church is the mechanism that God and the Spirit have used the past 1600 years to bring His message to each generation. Canonization was an act of God and the Spirit. It would be like God to have the Spirit based exegesis lose the battle to become the orthodox exegesis only to have the Spirit lead in the establishment of the orthodox Church.

The answer to the question of what to believe in, gnostic or orthodoxy, isn't either / or, but parts of both. It was both with Paul. The fundamental secret that the Gnostics had is "Christ in us". This is true. I have experienced numerous interventions in my life that I can only attribute to the work of the Spirit. However, the Biblical canon, not the Gnostic Gospels, is the mechanism that God has used to communicate with us both through the text and through spiritual insights consistent with the Biblical text. The cannon scriptures are the source of inspiration and the way to test spiritual insights.

The bottom line is that Paul's gospel is something new in the history of the world, the idea that God is in us. Through the death and resurrection of Jesus Christ, God has put His Spirit within those who believe in Him, and is thereby establishing a personal relationship with individuals. The Biblical text is the inspired word of God that ties all believers together. The Spirit is how God works with each of us individually to transform us according to His will.

Chapter 6. The Church

God speaks to us today through the Christian church and the church you attend.

In the previous chapters, we have used the term church, and said that Paul is credited with creating the Christian Church, as we know it today. But in Paul's day, there were no churches. Paul did not even consider himself to be a Christian. Here is what Dr. Wayne Meeks the eminent Pauline scholar says about Paul.

> "It (Paul's conversion) was not a change from Judaism to Christianity, which did not yet exist as a religion separate from Judaism, nor did it make Paul any less Jewish. Indeed, Acts portrays Paul as still a Pharisee many years afterward (23:6). He simply added to his existing Jewish beliefs his new conviction that the crucified and risen Jesus was indeed the Messiah, through whom God would fulfill all the promises he had made to Abraham. "[25]

Here is what God promised to Abraham:

> Genesis 17:6-7 (NASB)
> 6 "I will make you exceedingly fruitful, and I will make nations of you, and kings will come forth from you.
> 7 "I will establish My covenant between Me and you and your descendants after you throughout their generations for an everlasting covenant, **to be God to you and to your descendants after you.**

[25] *The Writtings of St. Paul, Page xxii,* Wayne A. Meeks & John T. Fitzgerald, W.W.Norton & Company New York 2007, 1972)

In Chapter 11 of Roman's, Paul describes how Gentiles could be grafted into Abraham's descendants through faith in Christ Jesus. God's promise is to be God to individuals from many nations for all generations.

Paul was called to bring the gospel to the Gentiles while Peter and the other disciples of Jesus focused on spreading the good news to the Jews. All of these men were Jews before Christ came into their life. All believed that Jesus was the Jewish Messiah. Christ's revelation to Paul was that there is a new covenant for all people based on faith. The new covenant focused on believing in Christ and receiving the Holy Spirit, not observing the Jewish laws of Moses. The other apostles wrestled with the issue of whether it was necessary to continue to follow the laws. This caused problems from the beginning when the new Jewish Christians came to Paul's new gentile Christian churches.

> Galatians 1:6-8 (NASB)
> 6 I am amazed that you are so quickly deserting Him who called you by the grace of Christ, for a different gospel;
> 7 **which is really not another**; only there are some who are disturbing you and want to distort the gospel of Christ.
> 8 But even if we, or an angel from heaven, should preach to you a gospel contrary to what we have preached to you, he is to be accursed!

When Paul says a different gospel, which is really not another, he means that it is the same good news concerning Jesus' resurrection, but a different way of achieving salvation, faith vs. following the Jewish laws. As we mentioned earlier, the church founded by Jesus' brother James eventually faded away because of its insistence on its members, including the gentiles, following Jewish laws.

In Roman times, some people were drawn to the idea of one god of the universe rather than worshiping the emperor and multiple gods. Groups of people welcomed Paul into their homes to hear of the new revolutionary ideas of Jesus' death and resurrection. Wealthy women led many of these groups. The groups that met in these people's homes became the churches that are described in the Bible in the latter part of the first and second century. Unfortunately, not much is known of the early churches. This is, in part, because Christians had to meet in secret to avoid persecution by both the Jews and the Romans. The persecution by the Jews stemmed from the fact that they had an established religion recognized by the Roman government. Christian beliefs not only challenged the Jewish faith, but also challenged the existing culture. Christianity was disruptive and dangerous.

As these Christian groups grew larger and started to correspond with one another, there needed to be some structure for the emerging religion. The different interpretations of Paul's teachings needed to be resolved and formalized. It also became clear that Jesus was not going to return as soon as Paul believed, so a way was needed to pass on the sacred texts. This is what prompted the writings of what later became the New Testament. Luke's Gospel and Book of Acts set the stage for the creation of the Christian church followed by the pastoral epistles attributed to Paul. Paul's gospel became the basic theology that underpins the church, as we know it today. However, as the church expanded and became more organized, there was a struggle between competing interpretations of Paul's message and other Christian beliefs. Historians believe that both Peter and Paul were brought to Rome and killed by the Roman government during Nero's reign as described previously. The Christian leaders that emerged felt that they could only survive if there was one unified belief called "Catholic" meaning "the whole" (katholike in Greek). They also looked for ways to legitimize their faith by linking it to an apostle of Christ.

By the second century, the church no longer focused on Paul as their founder (although they continued to follow his theology). The Bishops of Rome became the established orthodox church in the west, after many years of struggle, by claiming Peter as their founder. They based their legacy on Jesus' words in Matthew and John.

> Matthew 16:18-19 (NASB)
> 18 "I also say to you that you are Peter, and upon this rock I will build My church; and the gates of Hades will not overpower it.
> 19 "I will give you the keys of the kingdom of heaven; and whatever you bind on earth shall have been bound in heaven, and whatever you loose on earth shall have been loosed in heaven."

> John 21:16 (NASB)
> 16 He said to him again a second time, "Simon, son of John, do you love Me?" He said to Him, "Yes, Lord; You know that I love You." He said to him, "Shepherd My sheep."

The Catholic Church had begun with bishops presiding over church jurisdictions. Persecutions of the Christian churches continued, however, until 312CE when Constantine saw a cross in the sky and won a decisive military victory over Maxentius at Milvian Bridge. His troops brandished shields with a symbol of Christ during the final victorious battle. From that point on, the Roman Emperors embraced Christianity and established an alliance with the Christian bishops. That changed everything for the Christian church, and resulted in what we know as the Catholic Church today.

Here is how Dr. MacCullouch describes the end result:

> "Over the century and a half from Constantine's military victory in 312, emperors, armies, clergy, monks and excited mobs of ordinary Christians all contributed to a complex of decisions on which version of Christian doctrine was to capture the allegiance of the rulers of the world in the West and in Constantinople. The culmination of this process was a great council of Church leaders at Chalcedon in 451, under the control of the Roman emperor and his wife. We have already seen mainstream Christianity based on a series of exclusions and narrowing of options: Jewish Christians, Gnostics, Montanists, Monarchians were all declared outside the boundaries. Calcedon was to mark a new stage in this process of exclusion.....the new imperial church asserted itself as the one version of Christian truth for the world to follow, and, in the process, created a great deal of that truth for the first time."[26]
>
> "The copes, chasubles, mitres, maniples, fans, bells, censers of solemn ceremony throughout the Church from east to west were all borrowed from the daily observances of imperial and royal households. Anything less would have been a penny-pinching insult to God."[27]
>
> "Now, 'Catholic' Christianity was given monopoly status, not just against its own Christian rivals but against all traditional religion: ancient priesthoods lost all privileges and temples were ordered to be closed even in remote districts." [28]

---

[26] *Christianity The First Three Thousand Years*, Diarmaid MacCulloch, 2009, Viking Penguin, Page 190
[27] IBID Page 199
[28] IBID Page 220

While the Catholic Church became the monopoly religion, and the canonized Bible became the agreed to "word of God", the arguments did not subside. MacCulloch's 1016 page book captures the evolution of the church and significant church doctrine. However, even with a Pope to make decisions on behalf of God, differing Biblical interpretations and instructions to the people continued to grow around the world. As the Roman Empire broke up, the Christian Church in the East, now known as the Greek Orthodox Church, paid little attention to the Pope who resided in the West, in Rome. By the sixth century, the Church of the East was fully established in both theology and leadership and continued to expand. Islam emerged in CE622 disrupting the spread of Christianity in the East. Then, in the 1500s, came the Protestant Reformation in the West.

> "The Protestant Reformation, also known as the Protestant Revolt or the Reformation, was the European Christian reform movement that established Protestantism as a constituent branch of contemporary Christianity. It was led by Martin Luther, John Calvin and other early Protestants. The efforts of the self-described "reformers", who objected to ("protested") the doctrines, rituals and ecclesiastical structure of the Catholic Church, led to the creation of new national Protestant churches. The Catholics responded with a Counter-Reformation, led by the Jesuit order, which reclaimed large parts of Europe, such as Poland. In general, northern Europe, with the exception of Ireland and pockets of Britain, turned Protestant, and southern Europe remained Catholic, while fierce battles that turned into warfare took place in the central Europe. The largest of the new denominations were the Anglicans (based in England), the Lutherans (based in Germany and Scandinavia), and the Reformed churches (based in Germany, Switzerland, the Netherlands and Scotland). There were many smaller bodies as well. The most common dating begins in 1517

when Luther published *The Ninety-Five Theses*, and concludes in 1648 with the Treaty of Westphalia that ended years of European religious wars."[29]

The main-line churches that grew out of the reformation have continued to split and new denominations abound to this day.

If you look at the big picture of the history of the Christian church several things emerge.

1. There has been a concerted effort from the beginning to define the true meaning of Jesus Christ's life, death, and resurrection.
2. Since Christ's resurrection, people have argued over what we should believe and do to worship God. For example, one fundamental difference between Christian Catholics and Protestants is their primary focus. The Catholic faith focuses on Jesus' sacrifice and the sacrifice of saints throughout history. The Catholic Mass is derived from the "last supper", and their "stages of the cross" reflect Jesus suffering on the way to his crucifixion. Prior to the Counter-Reformation that began with the Council of Trent, the priest would celebrate the Mass on behalf of the people. Protestants' focus is on the "word of God" (the Biblical text), the resurrection, and the members of the congregation. Therefore, one might say that the Catholic Church is a place of God that contains the Host (body and blood of Christ), where Protestant churches are a place of God's people who look for God's revelations in the Bible.
3. People have struggled for over a thousand years to find the one true interpretation of the Bible.

---

[29] Euan Cameron, *The European Reformation (1991), (http://en.wikipedia.org/wiki/Protestant_Reformation)*

4. There is no definitive answer to 3 above. However, scripture says what it says whether you agree with it or not.
5. Power goes to those who get to do the interpreting of the Bible. Wars have been fought for that power.
6. Paul's letters and the Gospels are the closest text we have to answer 2 above. They are thick, layered, and conflicted.
7. The Catholic Church has been God's instrument to preserve the sacred Christian texts as the Jews have been for the OT.
8. The Christian Church has formed the basis of Christian belief. The Church we attend has formed many of our assumptions about God and the Bible.
9. And, the Church also taught us church doctrine that has also become a foundation of our faith. This has added to what we must do to please God if we believe that aspects of church doctrine are necessary to worship God.

However, the truth will set us free. God wants a personal relationship with us. The real role of the church is to help us establish a personal relationship with God- not build itself up. One of Paul's stories will help us understand:

> 1 Corinthians 3:1-23 (NASB)
> 1 And I, brethren, could not speak to you as to spiritual men, but as to men of flesh, as to infants in Christ.
> 2 I gave you milk to drink, not solid food; for you were not yet able to receive it. Indeed, even now you are not yet able,
> 3 for you are still fleshly. **For since there is jealousy and strife among you, are you not fleshly, and are you not walking like mere men?**
> 4 For when one says, "I am of Paul," and another, "I am of Apollos," are you not mere men?

5 What then is Apollos? And what is Paul? Servants through whom you believed, even as the Lord gave opportunity to each one.
6 I planted, Apollos watered, but God was causing the growth.
7 So then neither the one who plants nor the one who waters is anything, but **God who causes the growth.**
8 Now he who plants and he who waters are one; but each will receive his own reward according to his own labor.
9 For we are God's fellow workers; **you** are God's field, God's building**.**
10 According to the grace of God which was given to me, like a wise master builder I laid a foundation, and another is building on it. But **each** man must be careful how he builds on it.
11 For no man can lay a foundation other than the one which is laid, which is Jesus Christ.
12 Now if any man builds on the foundation with gold, silver, precious stones, wood, hay, straw,
13 each man's work will become evident; for the day will show it because it is to be revealed with fire, and the fire itself will test the quality of each man's work.
14 If any man's work which he has built on it remains, he will receive a reward.
15 If any man's work is burned up, he will suffer loss; but he himself will be saved, yet so as through fire.
16 **Do you not know that you are a temple of God and that the Spirit of God dwells in you?**
17 If any man destroys the temple of God, God will destroy him, for the temple of God is holy, and that is what you are.
18 Let no man deceive himself. If any man among you thinks that he is wise in this age, he must become foolish, so that he may become wise.

> 19 For the wisdom of this world is foolishness before God. For it is written, "He is THE ONE WHO CATCHES THE WISE IN THEIR CRAFTINESS";
> 20 and again, "THE LORD KNOWS THE REASONINGS of the wise, THAT THEY ARE USELESS."
> 21 So then let no one boast in men. For all things belong to you,
> 22 whether Paul or Apollos or Cephas or the world or life or death or things present or things to come; **all things belong to you,**
> **23 and you belong to Christ; and Christ belongs to God.**

This scripture is both pastoral and spiritual. The pastoral message is direction to the church members of Corinth to focus on God not on his messenger. One of the primary messages of 1 Corinthians is that all believers are one in Jesus Christ. Paul tries to teach the members of the church that they should focus on God and not the person sending them God's gospel. In a sense, he is saying the opposite of "Don't shoot the messenger". He is saying "Don't idolize the messenger"...use the message to come to a greater understanding of God. I think he is also saying that arguing over the messenger shows that you are not looking for the Spiritual insight that the messenger brings as described in 1Cor 2:10 "for God has revealed them to us by his Spirit".

So fast-forward to today. There are still wars fought by Christians against Christians of different beliefs as was the case in Northern Ireland. Look at how many denominations there are within the Christian church and how many people focus their attention on their priest, minister, or pastor (we can't even agree on what to call the church leader). Paul's pastoral message is as vitally important today as it was in Paul's day. I have seen this personally as part of the experience of writing my book on Revelation. We will talk more about Revelation later, but I will relate this story as an up to date example.

I have a friend who I will call Larry in this true story. Larry and I worked in the Pentagon together and soon we discovered that we were both Christians. We began having lunch together and would take the opportunity to discuss how God was manifesting Himself in our lives. It was great to have a friend whom I could talk to openly and receptively about the Bible and the work of God's Spirit. As time went on, however, I discovered that Larry had never really read the Bible, just certain verses that were associated with a particular church sermon. His thoughts and knowledge of the Bible came from his pastor who led a 2,000 member evangelical congregation. When I finished one of the final drafts of my book, I gave it to him for him to read and provide me with comments. Weeks later we met for him to provide me with feedback. What he told me surprised me. He told me that he had read my book four times, and each time he learned more. However, he didn't know what to say to me. He admitted that he didn't know enough about the Bible to comment on what I had written. Then the surprise came. He told me that his pastor had preached a sermon on Revelation that contained a different interpretation than the ideas described in my book. He said that he felt that he either needed to believe his Pastor or me. As a result he told me that I must be wrong. Well, I might be wrong. But, my reason for writing the book, aside from feeling compelled to do so, was to get people thinking, to test the views that they have held by digging into the Bible for themselves, and to see what the Spirit would show them. I reminded him of Jesus' words in Mark 8:29, "But who do <u>you</u> say that I am?"

Paul and Jesus' message to churches is that the members of the congregation are to establish a relationship with God, not focus on a spiritual leader or simply adopt someone's point of view. The church's role is to announce the gospel and provide spiritual food and support to the congregation, as Paul and Apollos did in Corinth. Your role in going to church is to worship God and to work to transform yourself according to God's will with the help of the clergy and the members of the congregation who are on their own journeys toward God. This is why it is so important that you surround yourself with a circle of friends and find a good church home where you can help each other grow in faith and service to God.

Chapter 7. John

God speaks to us today through the writings of the Apostle John.

There appear to be only two authors of the Bible who were personally taught by Jesus Christ, Paul and John. All the other authors wrote after the contemporaries of Jesus and Paul had died[30]. Paul's encounter with Christ was after He was raised from the dead, where John knew Him both before and after His crucifixion. John describes himself as the disciple who Jesus loved. He does this to show his closeness to Jesus and to humbly focus on Jesus not on himself.

Imagine being a close friend of Jesus while he was alive on earth. He chose you. You walk, talk, and eat your meals with Him. You witness the miracles that He is performing and hear Him speak his parables. At night, when you are lying on the ground looking up at the stars, you ask him the questions that are on your mind. And when you become old, and your friend has been crucified, and you see Him after He has arisen from the dead, you dictate your Gospel to the scribe. John's Gospel contains the messages he chooses to write. This closeness makes his Gospel especially important in understanding God's central messages to us. It also tells us how God communicated with John in his day, and how He continues to speak to us today. Jesus loves us as he loved John if we seek to come close to Him.

The key to John's Gospel is Chapter 20 verse 30 and 31. This is where he tells us what he has written based on all that he saw, did and heard during his life with Jesus.

---

[30] For information on what is known of the authors of the other NT books consult the *Harper Collins Study Bible,* New Revised Standard Version Bible, Harper One, copyright 1989.

> John 20:30-31 (NASB)
> 30 Therefore many other signs Jesus also performed in the presence of the disciples, which are not written in this book;
> 31 but these have been written so that you may believe that Jesus is the Christ, the Son of God; and that believing you may have life in His name.

When we read John's Gospel, we see that it is filled with quotes from Jesus that the Holy Spirit helped him to recall and write down. You should read John's entire Gospel. The following verses summarize John's Gospel with respect to how God communicates with us.

> John 1:1-5 (NASB)
> 1 In the beginning was the Word, and the Word was with God, and the Word was God.
> 2 He was in the beginning with God.
> 3 All things came into being through Him, and apart from Him nothing came into being that has come into being.
> 4 In Him was life, and the life was the Light of men.
> 5 The Light shines in the darkness, and the darkness did not comprehend it.

"The Word" is Jesus. The Word became scripture in the Bible. How does the Word communicate with us? John the Baptist was told that the way to recognize Jesus was through the Holy Spirit. This is the fundamental way God communicates with people on the earth and is the beginning of John's Gospel as follows.

> John 1:33-34 (NASB)
> 33 "I did not recognize Him, but He who sent me to baptize in water said to me, 'He upon whom you see the Spirit descending and remaining upon Him, this is the One who baptizes in the Holy Spirit.'

34 "I myself have seen, and have testified that this is the Son of God."

The following quote illustrates Jesus' use of symbols and parables when he talked to people. This quote shows that even his disciples didn't always understand what he was saying until after his resurrection.

John 2:18-22 (NASB)
18 The Jews then said to Him, "What sign do You show us as your authority for doing these things?"
19 Jesus answered them, "Destroy this temple, and in three days I will raise it up."
20 The Jews then said, "It took forty-six years to build this temple, and will You raise it up in three days?"
21 But He was speaking of the temple of His body.
22 So when He was raised from the dead, His disciples remembered that He said this; and they believed the Scripture and **the word** which Jesus had spoken.

Next we see that Jesus is telling us that the spirit is key to worshiping God. This is how we are to speak to God as well. Our spirit joins with the Holy Spirit to communicate with God.

John 4:23-24 (NASB)
23 "But an hour is coming, and now is, when the true worshipers will worship the Father in spirit and truth; for such people the Father seeks to be His worshipers.
24 "God is spirit, and those who worship Him must worship in spirit and truth."

Jesus focused on the Holy Spirit as the long-term messenger when He could no longer be with us on earth. His metaphor for the Spirit is "living waters".

John 7:38-39 (NASB)
38 "He who believes in Me, as the Scripture said, 'From his innermost being will flow rivers of living water.'"
39 Buy this He spoke of the Spirit, whom those who believed in Him were to receive; for the Spirit was not yet given, because Jesus was not yet glorified.

As we mature and grow in the Spirit, we learn to live our life as God leads us, as we see in Jesus' example. This may not be the easy road or what we would like to happen, but, as we come closer to God, pleasing Him becomes our greatest desire.

John 8:28-29 (NASB)
28 So Jesus said, "When you lift up the Son of Man, then you will know that I am He, and I do nothing on My own initiative, but I speak these things as the Father taught Me.
29 "And He who sent Me is with Me; He has not left Me alone, for I always do the things that are pleasing to Him."

Many people wonder what happens to people that die who are not Christians but appear to be good people. The following quote from John's Gospel may give us a clue. Only God knows to whom He is referring. We do know from Revelation 7:4-8 and 14:1-5 that Christ calls 144,000 from the tribes of Israel to heaven to be with Him during the end-times. They are not Christians, but they are delighted to learn the truth even though they do not believe Christ to be their messiah until He calls and seals them. Only God knows who else Christ will call and when. We are not to be judgmental. Our job is our personal relationship with God, as part of His flock not to condemn other people who do not believe as we do. But notice that the only way to be saved is for Christ to call you. He is the only way, but some do not realize this until they hear His voice. Christ's call is evidently more about one's spiritual connection to Him than one's theology.

John 10:14-16 (NASB)
14 "I am the good shepherd, and I know My own and My own know Me,
15 even as the Father knows Me and I know the Father; and I lay down My life for the sheep.
16 "**I have other sheep, which are not of this fold**; I must bring them also, and **they will hear My voice**; and they will become one flock with one shepherd.

John 5:22-29 (NASB)
22 "For not even the Father judges anyone, but He has given all judgment to the Son,
23 so that all will honor the Son even as they honor the Father. He who does not honor the Son does not honor the Father who sent Him.
24 "Truly, truly, I say to you, he who hears My word, and believes Him who sent Me, has eternal life, and does not come into judgment, but has passed out of death into life.
25 "Truly, truly, I say to you, an hour is coming and now is, when the dead **will hear the voice of the Son of God**, and ***those who hear*** will live.
26 "For just as the Father has life in Himself, even so He gave to the Son also to have life in Himself;
27 and He gave Him authority to execute judgment, because He is the Son of Man.
28 "Do not marvel at this; for an hour is coming, in which all who are in the tombs **will hear His voice**,
29 and will come forth; those who did the good deeds to a resurrection of life, those who committed the evil deeds to a resurrection of judgment.

In the following scripture, Jesus prays out loud so that those around him can see what has already happened in his spiritual communication with God.

John 11:41-42 (NASB)
41 So they removed the stone. Then Jesus raised His eyes, and said, "Father, I thank You that You have heard Me.
42 "I knew that You **always** hear Me; but because of the people standing around I said it, so that they may believe that You sent Me."

If you read the full Gospel of John, you will see that there are many, many quotes like the following. He repeats this so many times because this is the message that Jesus was trying so hard for people to understand. The Holy Spirit is the direct link between Jesus and God. The Holy Spirit made them one. We can be one with God as well if we let God's Spirit join with us.

John 14:10-11 (NASB)
10 "Do you not believe that I am in the Father, and the Father is in Me? The words that I say to you I do not speak on My own initiative, but the Father abiding in Me does His works.
11 "Believe Me that I am in the Father and the Father is in Me; otherwise believe because of the works themselves.

He goes on to say:

John 14:16-17 (NASB)
16 "I will ask the Father, and He will give you another Helper, that He may be with you forever;
17 that is the **Spirit** of truth, whom the world cannot receive, because it does not see Him or know Him, but you know Him because He abides with you and will be **in you**.

John 14:25-26 (NASB)
25 "These things I have spoken to you while abiding with you.

26 "**But the Helper, the Holy Spirit, whom the Father will send in My name, He will teach you all things, and bring to your remembrance all that I said to you.**

Here it is said plainly. The Spirit is responsible for giving John the remembrance of what Jesus said while He was on the earth in order for John to write his Gospel. And, the Spirit is sent from God in Jesus' name. So, Jesus' quotes in John are what God wanted to be preserved for us. The things Jesus wanted to say to us while he was on the earth are, therefore, written in the Gospel of John. Jesus Christ is the Word. Therefore, the Spirit will teach us all things from the Word of God as he did for John.

As Jesus was nearing his crucifixion, he spoke of the Spirit again and told his disciples to expect the Holy Spirit to come to them after his resurrection and give them greater understanding.

John 16:12-15 (NASB)
12 "I have many more things to say to you, but you cannot bear them now.
13 "But when He, the Spirit of truth, comes, He will guide you into all the truth; for He will not speak on His own initiative, but whatever He hears, He will speak; and He will disclose to you what is to come.
14 "He will glorify Me, for He will take of Mine and will disclose it to you.
15 "All things that the Father has are Mine; therefore I said that He takes of Mine and will disclose it to you.

After Jesus' resurrection, he appeared to the disciples, showed them his body, and then gave them the Holy Spirit as he said he would.

John 20:21-22 (NASB)
21 So Jesus said to them again, "Peace be with you; as the Father has sent Me, I also send you."

> 22 And when He had said this, He breathed on them and said to them, "Receive the Holy Spirit.

Finally, John says that this is a firsthand account from the disciple who Jesus loved. He then stakes his life and reputation that what he has written in his Gospel is true, and he reaffirms that what he has written is what we need to know. If this was not true his followers would have dismissed his writings after his death. Little did he know the number of books that would be written about Jesus Christ.

> John 21:24-25 (NASB)
> 24 This is the disciple who is testifying to these things and wrote these things, and we know that his testimony is true.
> 25 And there are also many other things which Jesus did, which if they *were written in detail, I suppose that even the world itself *would not contain the books that *would be written.

If we stand back from the quotes that we have just read, we can see something more. In each case, Jesus and John are speaking of individual relationships between God, the Spirit, the Bible / scriptures / the Word, and an individual. Each disciple received the Holy Spirit. God, the Father, is in Christ. Christ is in you. He is not talking about masses of people performing rituals, chanting established prayers, or offering sacrifices. The stories of miracles that are described between these verses are focused on individuals, raising Lazarus, talking to Nicodemus, and visiting at the well with the woman from Samaria.

John 20:24-29 is a story that specifically drives the point home that Jesus cares about each one of us. When Jesus appeared to his disciples after his resurrection, his disciple, Thomas, was not with the others, so he did not see Christ nor did he receive the Holy Spirit.

John 20:24-29 (NASB)
24 But Thomas, one of the twelve, called Didymus, was not with them when Jesus came.
25 So the other disciples were saying to him, "We have seen the Lord!" But he said to them, "Unless I see in His hands the imprint of the nails, and put my finger into the place of the nails, and put my hand into His side, I will not believe."
26 After eight days His disciples were again inside, and Thomas with them. Jesus came, the doors having been shut, and stood in their midst and said, "Peace be with you."
27 Then He said to Thomas, "Reach here with your finger, and see My hands; and reach here your hand and put it into My side; and do not be unbelieving, but believing."
28 Thomas answered and said to Him, "My Lord and my God!"
29 Jesus said to him, "Because you have seen Me, have you believed? Blessed are they who did not see, and yet believed."

On the surface, these short verses describe the reaction of one of Jesus' close disciples to reports that Jesus had risen from the dead, visited the disciples when Thomas was missing, and gave them the Holy Spirit. Thomas (who was called the twin and is now called doubting Thomas) was with the other disciples a week after the first event when Jesus appeared to them again. Jesus specifically returned to respond to Thomas' challenge and to tell him, you, and me that we will be blessed if we believe without such a demonstration. If you think about this for a moment, you can see that Jesus knew what Thomas was saying even when He was not with them, and that He was concerned enough about Thomas to return for his sake, for one person's sake. His Spirit is still doing this today.

One way to have greater insight into a section of scripture is to look at the context of the verses, i.e. the verses preceding and following the verses in question. In the case of the story of Thomas, the preceding verses show that this is part of John's description of what happened after Jesus' resurrection. He appeared first to Mary Magdalene, then to the disciples as a group, without Thomas, and then to the group including Thomas. If we look at the verses after those in question, we find John's description of the purpose of his Gospel "so that **you** may come to believe". This is at the end of his Gospel. From this, we can see that these verses about Thomas are the culmination of John's message to us. They are the last things that John describes. This shows how important this message is to John, the lasting impression. He is basically saying that God is a personal God who cares about each of us, wants us to come to him, and is involved in our life if we believe in him. We can learn from others in our community, and need to provide service with them and to them, but in the end our relationship with God is personal.

I believe that God's focus on the individual may also help explain why there is evil in the world. Many people say that there cannot be a loving god if He would permit world wars, holocausts, and other acts of evil. Here is the answer that I propose. God created the universe, men and women according to his laws and set it in motion. People are given free will to love God or reject Him. I describe this in more detail in the Epilogue and Appendix 2. God then focused on individuals. He speaks to individuals through the Bible, the work of the Holy Spirit, and the people and events around us. The evil in the world is caused by people who do not love God and are not guided by God's Spirit. God works to transform each individual who does love Him according to His will within this world that contains evil, and we are all on separate paths. His focus is on you, not what the world brings your way. I know that this sounds crazy, but God must allow evil in the world in order for people to have the option to love Him. Without free choice there is no true love. With free choice some people will do evil.

Chapter 8. Revelation

God speaks to us today through the Revelation to John.

When it comes to God talking directly to us, the Book of Revelation is unique. Revelation is the only book in the Bible where God specifically addresses us. Here are the opening verses.

> Revelation 1:1-7 (NASB)
> 1 The Revelation of Jesus Christ, **which God gave Him to show to His bond-servants**, the things which must **soon take place**; and He sent and communicated it by His angel to His bond-servant John,
> 2 who testified to the word of God and to the testimony of Jesus Christ, even to all that he **saw**.
> 3 Blessed is he who reads and those who hear the words of the prophecy, and heed the things which are written in it; for the time is near.
> 4 John to the seven churches that are in Asia: Grace to you and peace, from Him who is and who was and who is to come, and from the seven Spirits who are before His throne,
> 5 and from Jesus Christ, the faithful witness, the firstborn of the dead, and the ruler of the kings of the earth. To Him who loves us and released us from our sins by His blood—
> 6 and **He has made us to be a kingdom, priests to His God and Father**—to Him be the glory and the dominion forever and ever. Amen.

The idea that Christ knew what would happen in the end-times, but not the timing is plainly stated in the Gospels. Here is what Matthew has to say.

> Matthew 24:32-36 (NASB)
> 32 "Now learn the parable from the fig tree: when its branch has already become tender and puts forth its leaves, you know that summer is near;
> 33 so, you too, **when you see all these things**, recognize that He is near, right at the door.
> 34 "Truly I say to you, **this generation** will not pass away until all these things take place.
> 35 "Heaven and earth will pass away, but My words will not pass away.
> 36 "But of that day and hour no one knows, not even the angels of heaven, **nor the Son, but the Father alone**.

God tells us in Revelation He is revealing to Christ, John, and us what will happen shortly in this book. He is also saying that the generation to whom he is speaking will see all the things he is describing come to pass. But why shortly when the events described have not been fulfilled for almost 2000 years? Perhaps the generation to whom he is speaking is a future generation not the people of His day.

As we described in Chapter 3, God speaks in parables so that only the people he chooses, when he chooses, will understand. The book of Revelation (real name "The Revelation **to** John") is the ultimate parable. It is the only book that says not to add or delete from the words in the text. Revelation is also the only book in the Bible that promises a blessing to anyone who reads or hears it read. And, it has been the most widely interpreted of all books in the Bible. If there is one part of the Bible where we need the Holy Spirit to understand it's meaning, it is the book of Revelation. Yet, with the help of the Spirit, one can understand the mysteries of life through this book. I am living proof.

We have been talking about how the Spirit and the Bible work together to help us understand what God is trying to tell us. In 2003, I had a revelation about Revelation that resulted in my writing a book on Revelation titled "Jesus Reveals Revelation".[31] I will not attempt to describe all the insights that come from the book of Revelation in this book, but will use my experience in writing it as an example of how God spoke to me. I will then discuss God's Executive Summary and John's Preface to Revelation.

Most Biblical scholars do not believe that Revelation is a prophecy. They have studied first century Christian churches as best as they can, and the Roman Empire that were the rulers at the time. They conclude that Revelation is a coded message from John to the seven churches. In most scholars' view, John, not the apostle, was writing to the churches of his day to tell them to keep the faith during the terrible tribulations during the Roman Emperor Nero's reign. They believe that John wrote to the churches using symbology from the OT prophets. In this way, John could send the churches encouragement in terms that they were familiar with from reading OT scripture. Scholars believe that the Romans would not understand John's letter so it would not add justification to their persecution of the churches. I suggest that you read "The Theology of The Book of Revelation" by Dr. Richard Bauckham [32] and Dr. Elaine Pagel's "Revelations: Visions, Prophecy, and Politics in the Book of Revelation"[33] for a comprehensive discussion of scholarly views of Revelation.

However, my spirit would not rest with the answers of scholars. To begin, some scholars and I believe that it was the Apostle John that wrote Revelation. John was the disciple who Jesus loved and was not captured and killed during Nero's attack on

---

[31] *Jesus Reveals Revelation,* Charles H. Huettner, Booklocker, 2007 – 2010, www.JesusRevealsRevelation.com.

[32] *The Theology of The Book of Revelation,* Richard Bauckham, Cambridge University Press,1993 - 2008

[33] *Revelations: Visions, Prophecy, and Politics in the Book of Revelation*, Elaine Pagels, Viking Press, 2012

Christians, as were Peter and Paul. In addition, I believe that John's Gospel gives us a clue to the fact that he lived longer than expected to complete the work that Christ had for him to do.

> John 21:21-23 (NASB)
> 21 So Peter seeing him *said to Jesus, "Lord, and what about this man?"
> 22 Jesus *said to him, "**If I want him to remain until I come**, what is that to you? You follow Me!"
> 23 Therefore this saying went out among the brethren that that disciple would not die; yet Jesus did not say to him that he would not die, but only, "If I want him to remain until I come, what is that to you?"

I believe that John did remain until Christ came. He lived to see Christ come in the revelation that Christ showed His friend the Apostle John. John lived long enough to write the story of Christ's second coming for us.

While John certainly wrote to encourage the people of his day, I believe God was also speaking to another generation. Revelation has two messages; one that John was writing to his generation and a second that God made sure reached the last generation. The scripture itself tells us plainly its purpose as described above. Its length speaks to more meaning than simply encouragement to these seven churches at that time in history alone. The fact that some of the symbols are similar to those described by Daniel, Ezekiel, and other OT prophets could mean that those prophets saw some of the same visions as John. While Revelation certainly was written and sent to the seven churches with encouragement for their day and ours, I couldn't get over the feeling (and I know that millions of others feel the same) that there is much more that God is telling us in His prophecy.

As I pondered the book of Revelation, it seemed incomprehensible. Revelation appears to be a prophecy written to tell Christians throughout the centuries what will happen in advance of the end-times so that we will be prepared and keep the faith. It seems to be written chronologically from the time of John forward to the next world. Yet the stories of the harlot, various beasts, plagues, etc. do not tell a smooth flowing, understandable story. Many verses start with "Then" or contain "and then" that seems to indicate that the next events followed the preceding. But, the text does not tell a clear story that meets God's stated objective. This contradiction between a book to tell Christians what will happen, yet presenting a book that is not clear, seems to be a dilemma. It is a true parable. People for thousands of years have read the words, but don't understand without God's interpretation.

One night while I lay in bed, my mind wrestled with how to understand these things. My spirit tingled as I read the admonition to not change the "words" as written in Revelation 22:19. Why focus on words? That got me thinking about what could be changed that would provide me greater insight if I could not change the words. As I read Jesus words about the end-times in the Gospels, the answer became clear. The words in Revelation are true, but the order of events presented there are not. It appears to be written in chronological order, but it is not. This is a way that God could give all mankind the same words to read over the ages, yet seal up Revelation's true meaning until the end-time. The key to understanding Revelation is to reorder it to correspond with the order of events described by Jesus in the Gospels, Matt 24:9-34, Mark 13: 9-31, and Luke 21:7-33. Jesus words in these Gospels tell us how to interpret Revelation. This sent tingles down my spine.

But, this led me to three questions. First, why would John write Revelation out of order? The answer to this question is that he didn't. He wrote it in the order of the visions that God showed him.

> Rev 1:10 – 11 "I was in the Spirit on the Lord's day, and I heard behind me a loud voice like {the sound} of a trumpet, saying, **'Write in a book what you see,** and send {it} to the seven churches"

But God did not show John the vision in the order in which the events described would take place, thereby sealing it up until the end-times. Revelation is, therefore, the chronological story that John saw as scenes in heaven, scenes on earth, flashbacks, and prophecies. The word "then" means what John saw next not necessarily what would happen next. John's experience is like watching a play where you see all the acts and scenes of the play, but not in order and then you are told to write what you saw. It is only by reordering these scenes that the true chronology emerges. I believe that God did not want the true chronology to be known until the end-times, so that each generation would believe and look to the day of Christ's return even if it was actually to be hundreds of years in the future. As we enter the end-times, he wants us to know the straight story.

The second question is, if Revelation is out of order, how could we possibly know what order to put it in? As I began my efforts to discover how to reorder the text, I wanted to be sure that I was not reordering it to fit some preconceived idea as to what Revelation has to say. I worked under the assumption that if a scene didn't fit, that it was I who did not understand. For that reason, I began by using the original order unless there was a reason to change it and to use Jesus' words in the Gospels and the words in the text to signal what should be together. Here is an example of a time tag in the Gospels and in Revelation that shows how events can be synchronized:

Mark 13:24-25 (NASB)
24 "But in those days, after **that** tribulation, THE **SUN** WILL BE DARKENED AND THE **MOON** WILL NOT GIVE ITS LIGHT,
25 AND THE **STARS WILL BE FALLING** from heaven, and the powers that are in the heavens will be shaken.

Revelation 6:12-13 (NASB)
12 I looked when He broke the sixth seal, and there was a great earthquake; and the **sun** became black as sackcloth made of hair, and the whole **moon** became like blood;
13 and the **stars of the sky fell** to the earth, as a fig tree casts its unripe figs when shaken by a great wind.

Both Mark and Revelation are describing the same events. The reordered chapters came together in a way that was consistent with the Gospels. And, this event has not happened in the past so this must be a prophecy for the future.

The third question was what did God mean when he said in Rev 1:1 that the events would happen in a short period of time? This has been a problem for scholars and readers alike since it was written. This has led scholars to believe that they must look at events in the first century to explain Revelation. This is also why every generation has believed that their generation will see the end of times. But these events did not happen in a short period of time from when Revelation was written. Is the revelation wrong, or is there another interpretation?

The text says that the revelation is to show His bondservants / followers / believers, but it does not say which ones. Most assume it means the churches to which John is writing. Another interpretation could be that it is addressed to the end-time bond-servants that see the things that Jesus is describing in Matt 24:9-34, Mark 13: 9-31, and Luke 21:7-33. The letters were sent to the seven churches to make sure that the prophecy was preserved. It became clear that Jesus is the key to understanding to whom the revelation is addressed. I believe that **God's** real audience is not the disciples Jesus is talking to in the Gospels, or the churches that John is writing to in Revelation, but the bondservants that see the signs that Jesus describes in the Gospels.

> Mark 13:29-30 (NASB)
> 29 "Even so, you too, when **<u>you</u> see these things happening**, recognize that He is near, right at the door.
> 30 "Truly I say to you, **this generation** will not pass away until all these things take place.

The literal reading of the revelation has helped all generations to get a glimpse of the end of the Biblical story and be assured that Christ will return in triumph, but the details of this parable have been sealed until the end-time generation.

Another revelation to me was that these three descriptions by Jesus to his disciples all say that "this generation" will see the end of it all. The word "this" is singular. This means that the bondservants he is speaking to "will see the end of it all" **within one generation.** The end-time generation will see all that Jesus describes in the Gospels and the events described in Revelation.

As I gained more and more insights from the Spirit, the scriptures began to open up and I received greater understanding. Each subsequent time through, the text and the Spirit opened more doors until the order and the meaning became clear. What emerged over many drafts is a chronology with distinct subdivisions that can be seen in the text based on God's involvement with his chosen people and recipients of His covenants, the Jews and the Christians. I also began to realize that Revelation not only tells the story of the end-times, but also why we are born, why good people suffer, the completion of God's covenant with the Jews as well as the Christians, and many more things. It is the parable that makes clear many of the other parables in the Bible. For example it answers the mystery described by Paul.

> 1 Corinthians 15:51-52 (NASB)
> 51 Behold, I tell you a mystery; we will not all sleep, but we will all be changed,
> 52 in a moment, in the twinkling of an eye, **at the last trumpet**; for the trumpet will sound, and the dead will be raised imperishable, and we will be changed.

When Revelation is reordered, we find that the last trumpet (trumpet 7) does signal the return of Christ. This occurs in Chapter 10 of Revelation, but when it is reordered according to Jesus' words, it occurs late in the story as Christ returns and the Christians alive during the tribulation are raised. Here is what it says:

> Revelation 10:5-7 (NASB)
> 5 Then the angel whom I saw standing on the sea and on the land lifted up his right hand to heaven,
> 6 and swore by Him who lives forever and ever, WHO CREATED HEAVEN AND THE THINGS IN IT, AND THE EARTH AND THE THINGS IN IT, AND THE SEA AND THE THINGS IN IT, that there will be delay no longer,
> 7 but in the days of the voice of the **seventh angel, when he is about to sound (*trumpet 7*),** then the mystery of God is finished, as He preached to His servants the prophets.

I can't prove that the interpretation that resulted from my spiritual journey is the correct interpretation. Nor can anyone say definitively that it is wrong. The point for this discussion is that I know that the Holy Spirit guided me. I saw no visions. But I did receive insights about the Bible that gave me numerous "Ah-Ha" moments and a burning desire to share what I had learned with others. The results of my Spiritual insights and my study of the Bible resulted in an overwhelming urge to write my book on Revelation. Subsequent to that, it has opened doors for me to personally witness to hundreds of people about the power of God and His spirit. Beyond that, through an amazing series of "coincidences", it brought me to the Virginia Theological Seminary where I gained the understanding that led me to write this book. As we will discuss later, the way to know if the Spirit is working within you is by seeing that the resulting fruits of your labors are consistent with what we are told in the Bible. In this case, my efforts have produced fruit for the glory of God.

As for the book of Revelation itself, there is too much for me to address in this book, however, I will discuss Revelation's Executive Summary that was written by God and its Preface written by John. [34]

Revelation verses 1-8 are God's Executive Summary. Verse 1 describes how the revelation comes to John, and Verse 2 tells us why. Verse 3 echoes the blessing in the final verses of revelation. Verse 4 says that the book is written by John to the churches. Verse 5 says that John received the revelation from God, the Spirit, and Christ and describes Christ's credentials. It also says that Christ loves us and saved us by his death. Verse 6 tells us the end of the revelation story and our future as Christians; that we will be a kingdom and priests of God. Verse 7 tells the punch line of the revelation by describing how Christ will return and the fact that all those who ever lived will see him return including those who pierced Him as He hung on the cross. Verse 8 is God's signature block and completes the executive summary to the Revelation. See for yourself.

---

[34] *"Jesus Reveals Revelation"* Charles H. Huettner, Booklocker, 2007 – 2010, http://booklocker.com/books/3127.html, Page 20.

Rev 1:1-20
CHAPTER 1

1 The Revelation of Jesus Christ, which God gave Him to show to His bond-servants, the things which must shortly take place; and He sent and communicated {it} by His angel to His bond-servant John,
2 who bore witness to the word of God and to the testimony of Jesus Christ, {even} to all that he saw.

If you read these verses closely they say that this is the revelation that God gave to Jesus Christ so that Christ could show his bondservants what will happen. So this is actually God's revelation to Christ. And Christ sent it by his angel to John. God's plan was not only to inform Christ, but his bondservants as well. Who are Bond Servants? Revelation 11:18 defines bondservants as Prophets, Saints, and those who fear God's name. In this first verse of Revelation, God refers to John as a bondservant of Christ because he bore witness to Jesus. He also was referring to you and me, John's brothers and sisters, if we fear God and bear witness to Christ.

In addition, God gave Jesus this revelation to show His followers / bondservants "the things which must shortly take place...". This is why every generation has believed that their generation will see the end of times. As I discussed previously, an interpretation could be that it is particularly addressed to the end time bond-servants that see the things he is describing in Matt 24:9-34, Mark 13: 9-31, and Luke 21:7-33.

3 Blessed is he who reads and those who hear the **words** of the prophecy, and heed the things which are written in it; for the time is near.

Revelation 22:18 tell us of the same blessing. So, Revelation begins and ends with the same message. Even if you can't read, you can receive this blessing if you act on what it tells us. Once again, the focus is on words not the order of events.

> 4 John to the seven churches that are in Asia: Grace to you and peace, from Him who is and who was and who is to come; and from the seven Spirits who are before His throne;

Revelation refers in various places to churches and synagogues. Churches refer to the church of Christ (Christians, although they were not called this then). While synagogues refer to the covenant with Abraham, the Jews. This verse, therefore, tells us that Revelation is written to Christians, but Revelation tells us that it describes what will happen to fulfill God's covenant with the Jews as well.

Notice that "Him who is and who was and who is to come" is God the father not Jesus Christ. The seven Spirits are before the throne of God not Christ. The next verse, 5, says "and from Jesus Christ" not who is Jesus Christ.

> 5 and from Jesus Christ, the faithful witness, the first-born of the dead, and the ruler of the kings of the earth. To Him who loves us, and released us from our sins by His blood,

Verse 5 is the story of Jesus in two phrases. It is certainly an executive summary of who Jesus Christ is. How remarkable.

> 6 and He has made us {to be} a kingdom, priests to His God and Father; to Him {be} the glory and the dominion forever and ever. Amen.

How long has man wondered why we are here on earth? Here is the answer. Christ created us to be part of his kingdom and to be priests of God. This is the executive summary of who we are.

> 7 Behold, He is coming with the clouds, and every eye will see Him, even those who pierced Him; and all the tribes of the earth will mourn over Him. Even so. Amen.

This is a reference to Revelation 14:14 when Christ returns to earth. It also says that both the living and the dead, including those who crucified him, will see him when he returns to earth. The text ends with Amen to show that this is the end of what God has to say in his executive summary.

> 8 "I am the Alpha and the Omega," says the Lord God, "who is and who was and who is to come, the Almighty."

In this verse, God is telling us who he is. Interestingly, God told Moses that his name is “I Am”.

> Exod 3:13-15
> 13 Then Moses said to God, "Behold, I am going to the sons of Israel, and I shall say to them, 'The God of your fathers has sent me to you.' Now they may say to me, 'What is His name?' What shall I say to them?"
> 14 And God said to Moses, "I AM WHO I AM"; and He said, "Thus you shall say to the sons of Israel, 'I AM has sent me to you.'"

This is an illustration of how verses across the Bible can bring insight to what you are reading in Revelation. Once we know that God’s name is I AM, then we see that this is not only a statement from God, but also His signature block at the end of His Executive Summary. I would sign a document for my business:

Charles H. Huettner, President
Charles Huettner Associates, LLC

God signed His Executive summary and cover letter to the book of Revelation:

I AM, The Almighty
The Alpha and the Omega, who is and who was and who is to come.

I Am, (his name), The Almighty (His title), The Alpha and the Omega, who is and who was and who is to come (Who he is),

So, in these first 8 verses of Revelation we have the top-level version of the book of Revelation and the Bible authored by God[35]. This is the only portion of the Bible specifically dictated by God that includes his signature. If you only have time to read 8 verses of the Bible, this is what you should read and thoroughly understand. What follows in the first chapter of Revelation is the Preface of the book written by John as directed by Jesus Christ.

> 9 I, John, your brother and fellow partaker in the tribulation and kingdom and **perseverance** {which are} in Jesus, was on the island called Patmos, because of the word of God and the testimony of Jesus.

Verses 1 – 8 were dictated by God. From verse 9 on, John is the author telling the story of what he saw and has been told to write. The remainder of Chapter 1 of Revelation is John setting the scene for his writing of what he has seen. This is where John has the opportunity to speak to us from his experience as a prophet, disciple, and friend of Jesus Christ on earth. It is his attempt to put Jesus Christ and his revelation into prospective.
Verse 9 may also be one of the most important verses in the Bible for people who have suffered a tragedy in their lives. It tells us what it is to be "in Christ". It is odd that most people, including

---

[35] Isn't it amazing that after all the battles over the canonization of the Bible that Revelation would end up being the last book in the Bible and would begin with God's executive summary and signature block? If the Bible is divinely inspired, wouldn't God want to sign his work at the end then provide us with a epilogue filling in the pieces that are missing?

myself, seem to skip over the first portion of this verse and focus on John being on Patmos. Yet the first portion of the verse is key to understanding Revelation and your life. If we reorder Verse 9 it plainly says that to be in Christ we must partake in tribulation. If we persevere then we will be part of his kingdom. This is the story of Revelation in a nutshell, and it is John's first sentence in the preamble to his book. The remainder of the book is filling in some incredible details including what will happen as the end-times begin, the end of the covenant with the Jews, what happens after we die, and many other things.

With regard to your life, I am sure that you have had things happen that have caused you to wonder why a loving God would allow "this" to happen; perhaps the death of a loved one, or a tragic disease, injury or injustice. What this verse is telling us is that to be "in Christ" you will go through tough times, adversities, tribulations, and it is up to you to persevere to enter the kingdom. This was the story of Jesus' life here on earth. Why should we expect something different for we who are Christ's followers? Those that turn away from Christ when times are tough fail to persevere and will fail to be part of Christ's kingdom. The gospel is full of verses that tell this story, such as the seed that falls on the rocky path,[36] yet here it is spoken plainly as part of the introduction to Revelation.

The bottom line is that as far as God is concerned, the tribulation isn't the most important issue; it is how you respond to it that really counts.

---

[36] Matthew 13: 18-23

Chapter 9. Context

God speaks to you today through your daily life.

In day-to-day life, how can you know when the Holy Spirit is trying to guide you? How can you gain a better understanding about what the Biblical scriptures are saying?

Context is the key. There are two kinds of context that scholars are discussing today, the scriptural context that helps us understand the Bible, and the real world context that helps us see the work of the Spirit in our lives and others. Let's begin with the scriptural context.

Scriptural context is the idea that we can learn more about what Bible verses mean by viewing them in the context of the text before and after the verses in question. The objective here is to see a bigger picture of what the author, and perhaps God, is trying to say. This brings us above the specific words used to the broader meaning.

The first step in applying this concept is to disregard the Bible's chapter breaks and its numbering system. These did not occur in the original text, but have been added over the centuries to help readers. However, they often also break the authors messages in ways that lead to a more narrow and perhaps incorrect understanding. A good example of this is the story of "Doubting Thomas" that we discussed earlier.

John 20:24-29 describes what appears to be a self-contained story of Jesus proving to Thomas that He had risen from the dead. We previously said that John's telling this story is to show that Christ is concerned about each one of us. We could see this by reading the verses directly after the ones describing Thomas' meeting with the risen Christ. We can gain additional insight if we look to the verses prior to these, starting at John 19:31. We find that in John's description of Christ proving He had risen to Thomas, John is also trying to prove something else. The full story begins in Chapter 19, Verse 16. Read from there to John 20:29. Verse 19:34-35 says that Jesus' side was pierced, and that He (John) was **testifying to what he saw** so "you may also believe". In verse 20:27 the risen Christ tells Thomas to put his hand into his side and said "do not doubt, but believe". If we then read the verses beyond the story of Thomas you find that John says that his entire book is written, " so that you (the reader) may come to believe". Taken together we find that his reporting on Jesus being pierced is his way of validating that he, John, is a true, eyewitness reporter. Also, that Jesus piercing was done not only to fulfill prophecy, but also so that we would believe John and through John's book believe that "Jesus is the Messiah, the Son of God and that through believing you (we) may have life in His name". So the story has several meanings if we look beyond the short verses about Thomas. They are to prove that Christ has risen, that Christ is concerned about each one of us, and that John is an actual witness and a truthful reporter. Therefore, one way to gaining a deeper understanding of the Biblical text is to find the context of the scripture you are seeking to understand, and to look for the big picture of what is going on and being said. There may be more than one thing going on simultaneously, as in this case.

Next, let's turn to the Real World Context. In Chapter 4, we discussed how scholars have tried to find the true meaning of the Bible over the last 150 years by studying the history of the text and the time when it was written. Scholars today are beginning to realize that context is more than the cultural context of when the texts were written. Here is what Dr. Meeks has to say:

> Thus, by seeing the past as it really was, we thought to find a secure starting place for knowing and valuing. Instead, we found texts and lives that are always open to

> interpretation. We discovered meaning to be dependent always on context. We found identity to be a social process. And we learned that we could not avoid involving ourselves whenever we attempted to assess historical events and personages in any non-trivial way[37].

This leads us to a discussion of context within the real world. As Dr. Meeks said in the quote above, everyone who reads the Scriptures sees something different depending on the context of their lives and the lives of the community of which they are members. You can pick almost any verse in the Bible and there will be agreement about the interpretation within a church or certain group, but perhaps not across churches. Take the first verse in the Bible: Genesis 1:1 "In the beginning God created the heavens and the earth." and acknowledge the resulting controversy that goes back to Darwin. As scholars have come to the conclusion that there is no way of absolutely determining the "right" interpretation of specific Biblical texts other than the basic Christian concepts, they have begun to favor what is called Contextual Bible Study.

Dr. Gerald West's presentation at the 2011 Trinity Institute[38] begins to describe Contextual Bible Study as a combination of scriptural and real world contexts. As an example, he uses Mark 12:41-44 as the text to be studied.

> Mark 12:41-44 (NASB)
> 41 And He sat down opposite the treasury, and began observing how the people were putting money into the

---

[37] *The Writings of St. Paul,* Wayne A. Meeks and John T. Fitzgerald, 2007, W.W. Norton & Company, Inc., Page 692

[38] *Trinity Institute® January 19-21, 2011* Gerald West: Contextual Bible Study Presentation

treasury; and many rich people were putting in large sums.
42 A poor widow came and put in two small copper coins, which amount to a cent.
43 Calling His disciples to Him, He said to them, "Truly I say to you, this poor widow put in more than all the contributors to the treasury;
44 for they all put in out of their surplus, but she, out of her poverty, put in all she owned, all she had to live on."

This story is generally understood to focus on verse 44 as the meaning of this story. However, if you read the verses prior to this, you find a discussion about how the Jewish leaders were concerned more about themselves than the people they were supposed to be helping. Here are the previous verses:

Mark 12:38-40 (NASB)
38 In His teaching He was saying: "Beware of the scribes who like to walk around in long robes, and like respectful greetings in the market places,
39 and chief seats in the synagogues and places of honor at banquets,
40 who devour widows' houses, and for appearance's sake offer long prayers; these will receive greater condemnation."

A real revelation comes when you read the text after the description of the poor woman that appears in the next Chapter, Chapter 13. Here is what it says:

Mark 13:1-2 (NASB)
1 As He was going out of the temple, one of His disciples said to Him, "Teacher, behold what wonderful stones and what wonderful buildings!"

> 2 And Jesus said to him, "Do you see these great buildings? Not one stone will be left upon another which will not be torn down."

This text has been interpreted as Jesus predicting the destruction of the temple and the beginning of His discussions of the end-times. If we put these three sets of verses together, however, we can begin to see that it is not the physical temple that Jesus is talking about. He was talking about the destruction of his body, his crucifixion. However, beyond this, He may also be referring to the destruction of the Jewish law based religious system exemplified by the scribes in the first of the three sets of verses, to be replaced by what became known as Christianity after His resurrection. The widow's story in the broader context is part of Jesus foretelling the new covenant through his blood. By looking at all three sets of verses, the widow's story exemplifies the corruptness of the scribes and their cold-hearted religious system at that time that Jesus would overturn through his crucifixion and resurrection.

West goes on to suggest that having read the widow's story within the context of the verses before and after it, we should consider what it is saying within our own real world context. Would Jesus say the same thing about our churches? What does this say about our culture and political system?

Contextual Bible Study can be summarized by answering the following 5 questions.

1. What does the passage mean? – What does the passage say literally?
2. What does the passage mean in context? – Look at the verses before and after to identify the full story and see if that provides added meaning.
3. What does the passage mean to me? – Implies that the reader will understand the passage based on his / her background.

4. What does the passage mean to US? – Implies a community of belief such as a given church. Also implies a boundary between us and them (those who do not interpret it as "we" do).
5. Who am I? – One's personal context in reading this passage. Insight can be gained by exploring the differences between them and us.

What scholars are discovering is that there are communities of readers who determine the "right meaning" of the Scriptures. There are multiplicities of readers, each holding a meaning within a culture either intentionally or by osmosis. A poor black woman in Africa can read the Biblical texts and understand them differently than a wealthy white male living in the United States. A challenge that the Bible presents to us all is to begin to understand our own biases based on our background, community of belief, and economics, and be open to other peoples' understanding based on their context.

As we stand back from this discussion of context, we begin to see that church scholars are beginning to realize that the role of the church needs to be transformed from a place where people go to get finite answers to a place where the church can respond to the people and help them to address their needs and grow in faith. But does that mean that there aren't any answers except what a group of people agrees on? NO.

There is one other context that is not being discussed by the scholars. The context of the Holy Spirit working in our individual daily lives. Contextual Bible Study is a useful way to understand the Bible and those around you, but the real issue is living with Christ in you. Your personal relationship with God day to day is the ultimate objective. **I propose that God uses your daily life as a way to bring you closer to Him.**

1 Corinthians 2:1-16 (NASB)
1 And when I came to you, brethren, I did not come with superiority of speech or of wisdom, proclaiming to you the testimony of God.
2 For I determined to know nothing among you except Jesus Christ, and Him crucified.
3 I was with you in weakness and in fear and in much trembling,
4 and my message and **my preaching were not in persuasive words of wisdom, but in demonstration of the Spirit and of power,**
5 so that your faith would not rest on the wisdom of men, but on the power of God.
6 Yet we do speak wisdom among those who are mature; a wisdom, however, not of this age nor of the rulers of this age, who are passing away;
7 but we speak God's wisdom in a mystery, the hidden wisdom which God predestined before the ages to our glory;
8 the wisdom which none of the rulers of this age has understood; for if they had understood it they would not have crucified the Lord of glory;
9 but just as it is written, "THINGS WHICH EYE HAS NOT SEEN AND EAR HAS NOT HEARD, AND which HAVE NOT ENTERED THE HEART OF MAN, ALL THAT GOD HAS PREPARED FOR THOSE WHO LOVE HIM."
10 For to us God revealed them through the Spirit; for the Spirit searches all things, even the depths of God.
11 For who among men knows the thoughts of a man except the spirit of the man which is in him? Even so the thoughts of God no one knows except the Spirit of God.
12 Now we have received, not the spirit of the world, but the Spirit who is from God, so that we may know the things freely given to us by God,

> 13 which things we also speak, not in words taught by human wisdom, but in those **taught by the Spirit, combining spiritual thoughts with spiritual words.**
> 14 But a natural man does not accept the things of the Spirit of God, for they are foolishness to him; and he cannot understand them, because they are spiritually appraised.
> **15 But he who is spiritual appraises all things**, yet he himself is appraised by no one.
> 16 For WHO HAS KNOWN THE MIND OF THE LORD, THAT HE WILL INSTRUCT HIM? But we have the mind of Christ.

If we read this passage and pay particular attention to the sections I have highlighted, you can see that Paul is telling us that the Spirit teaches us by combining spiritual thoughts (from God's Spirit to ours) with spiritual words (the words in the Holy Bible). He then goes on to say that we who are filled with the Holy Spirit appraise all things. What this means is to look at the events in our lives as they transpire and search for the interventions of the Holy Spirit. How can you tell? They are the things that happen out of the ordinary that unsettle you spirit. When you see something unexpected in real life, stop long enough to think about why this might have happened. When you identify such an event, look for the context within which it occurred. See how that real life context suggests something you should read in the Bible. If you take some time each day to read your Bible, events of your life might bring certain verses to mind. Through prayer and study, the Spirit will speak to you. The first step, however, is to be listening and watching as you live your life. This will result in a day-to-day, moment-to-moment relationship between you and God that will grow as you learn how to listen to the Spirit and practice this approach. Hearing the theory doesn't work. Practicing it is essential.

What does this look like in real life? I can only tell you from my own life and hope that this gives you some insight. What I have found is that listening to the Spirit is not a single event as described in the interview of Francis Collins in Chapter 2. As you see the Spirit working in your life, your ability to listen increases and your faith grows to where you are expecting God's help in your life and seeing it happen.

I have had many personal experiences of the Spirit acting in my life, but I will tell you three interconnected stories relating to my career. It is through experiences like these that I grew in my ability to listen to the Spirit. Let's begin with a career decision I made many years ago where the Spirit guided me through events in my daily life.

Career Decision- I want to be an Airline Pilot: By the mid 1970's, I was off active duty and in the Air Force Reserves. I was an Air Force pilot with over 3,000 hours of heavy jet flying time, and my dream was to become an airline pilot. Unfortunately, none of the airlines were hiring, so I joined the FAA as an Air Carrier Operations Inspector. One day, I saw a notice that United Airlines would begin to hire its first pilot training class in 7 years. This was my chance. I called United's Chief Pilot at Washington National Airport and asked if I could meet with him. He agreed. This began one of my close encounters with the Holy Spirit.

On the day of my meeting, my wife picked me up at FAA headquarters and drove me to United's office in a hangar at National Airport. There was an empty parking slot near the front door, so we parked and I went in to talk to the Chief Pilot. I described my aviation career with the military and FAA. I also told him that I had received B-747 training at United's training center in Denver and had an Airline Transport Pilot rating as a result. This was music to his ears. I had experience in large jets, with the FAA, the agency that regulates United, and I had proven myself to United's own training department in their most complicated aircraft. He gave me an application, put his initials at the top, and invited me to send it back to him personally. I was on cloud nine. If I could get into United's first training class in years, at my age, I could one day become their Chief Pilot. This is because all the pilots senior to me would one day be retired before me.

My head was spinning as I left the building and got back into my car. I joyfully told my wife what had happened and went to start the car. The battery was dead. When I turned the key, I could not even hear a click. I couldn't believe it. We had just driven and parked 15 minutes before. As I was trying to figure out what to do, someone stopped and asked me if they could help me jump-start the car. I gratefully agreed. After the car started, my wife drove me back to work. We never had battery trouble in that car again. This, however, was only the first of a series of unusual events that followed the interview.

That evening I completed the United Airlines pilot application with great enthusiasm. At that point, I began to feel uncomfortable. I decided that I would wait until the next day and review the application again. Perhaps my uncomfortable feeling was because I had done something wrong on the application and I would discover what it was when I reviewed it with fresh eyes. The uneasiness continued the next day at work, so I was anxious to get back to the application when I got home. That evening, I reviewed the application again, but could not find any problems. So, I put it in the envelope, added several stamps so I surely had enough postage, and went out the front door to mail the application at the mailbox on the corner... I could not put the application into the mailbox.

As I opened the mailbox, my uneasiness turned to dread. That is the only way I can describe it. Like the feeling you have when you hear that a loved one has died. I backed away from the mailbox and brought the application back home. If I had known the things I am writing about in this book, I would have known that this was from the Holy Spirit, but I didn't understand at the time. I thought that I should simply give it some time.

I went to work the next day and received a call from my boss at the FAA. He told me that I had been selected to enter an executive training program that would lead to my becoming a member of the federal Senior Executive Service and a position of responsibility. This came out of the blue at the precise time I was wrestling with the decision to apply to United.

That night was a restless one for me. My dream had been to become an airline pilot. How could I simply walk away from this fabulous opportunity? Work only 15 days a month, much higher salary than FAA, free airline passes- "I am a sucker to turn this down". Yet something wasn't right. That brought me to the Bible and prayer. I don't remember what I read or said to God. There was no burning bush or vision, just a gentle calm as I decided to accept the FAA's offer. I sometimes think you don't know what the right answer is until after you make it. But, in this case I am sure that the Holy Spirit was guiding me.

As it turned out, many of my friends did become airline pilots. Under the FAA regulations, they were forced to retire when they became age 60. Their timing was such that they reached 60 in 2006. If you recall, the airlines were all in deep financial trouble at that time and had unfunded retirement programs. One of my closest friends had earned a $95,000 retirement that was slashed to $23,000. I, on the other hand, had many unbelievable opportunities to improve aviation safety, contribute to our country's aerospace industry, and retire from government with a great government pension. I certainly wouldn't have guessed that this would be the outcome in 1977. And, I was led by the Spirit along the way, as I will further describe.

AeroClub Luncheon. The experience just described helped me to learn how to sense the work of the Spirit in my daily life. This came home to me several years later at a lunch sponsored by the Washington AeroClub. The AeroClub is an organization with members from all segments of the aerospace community in Washington D.C. Members include D.C. representatives of all the major aerospace companies, airlines, airports, industry associations, lobbyists, consultants, and congressional aids. They have a luncheon each month at a D.C. hotel featuring a guest speaker. There are routinely several hundred people in attendance. One month, unexpectedly, I was asked to sit at the head table with the dignitaries. As it turned out, two FAA Administrators were also at the table, one seated on my left and one on my right. As part of the agenda, the current Administrator was presenting the preceding Administrator with his official portrait that was to hang in the FAA building.

As background to the punch line of this story, I need to say that several years prior to this, under the preceding Administrator's administration, the issue of infant child restraints had come to the foreground. I was the head of the FAA Safety Office, and a colleague was the head of the FAA's Regulatory Office. The issue came to a head at a meeting of the Administrator's Management Team that included all of our peers. I made the case that FAA requires everything to be restrained in the aircraft except babies under 2 years old, and that FAA should require some restraint, not simply let the child sit in the parent's lap where turbulence would send the child flying in the cabin. (A 20-pound baby weighs 60 pounds instantly in 3G turbulence.) My colleague made the case that the economic consequences of requiring babies to be restrained would result in people driving rather than flying. Since more people are killed on the road then in airplanes, this would result in less safety. My response was that we were responsible for aviation safety and should fulfill our duty, especially to those who could not defend themselves from turbulence and accidents, i.e. babies. All eyes went to the

Administrator who was the decision maker. What he said gutted the FAA Safety Office's authority in the Agency. He said that he didn't know how he could decide against the recommendation of the head of his regulatory organization without removing him from his position, and he was not going to do that. In other words, infants will not be required to be restrained (and as of this writing decades later, they still are not). The implication of this decision was that any time the Safety Office disagrees with the Regulatory Office, the Regulatory Office would win. The next Administrator had recently told me that I would not be selected as the new head of the Safety Office, and eventually he closed the Safety Office entirely. This brings me back to the story of the luncheon.

I was now sitting between the Administrator who had gutted my office and the Administrator who decided not to select me to run it. As the luncheon program began, the President of the Club, most amazingly, introduced me first. He asked me to stand up behind the head table, told those at the luncheon of my many successes for aviation safety, called me Mr. FAA Safety, and asked for a round of applause. While this was extremely flattering, what came to me was that I was living the 23rd Psalm.

> Psalms 23:4-6 (KJV)
> 4 Yea, though I walk through the valley of the shadow of death, I will fear no evil: for thou art with me; thy rod and thy staff they comfort me.
> 5 **Thou preparest a table before me in the presence of mine enemies: thou anointest my head with oil; my cup runneth over.**
> 6 Surely goodness and mercy shall follow me all the days of my life: and I will dwell in the house of the LORD for ever.

While these were not really my enemies, and my life was not in danger, the parallel to the 23rd Psalm struck me as I stood before the audience. I felt that the Holy Spirit was giving me the experience of this Bible parable first hand within the context of my life. It helped to build my faith and encouraged me to persevere even during this time of adversity. This is an example of the work of the Bible and the Spirit coming together in my daily life.

Passed Over for Promotion. Just prior to the luncheon that I just described, the new FAA Administrator called me into his office and told me that he wanted to put another person from outside of the agency in as the head of the Safety Office that I had been successfully running for most of the past six years. I realized that I would not get the promotion, that the new Assistant Administrator would be uncomfortable with me in the office as his Deputy, and that my opportunities for advancement at FAA were probably over. By this time, I was keenly aware of the Holy Spirit working in my life and I couldn't understand why this was happening to me. Surely God would not have guided me into working for FAA only to leave me stuck in a position where I couldn't be effective until the end of my federal government career.

And then a miracle occurred. As we left the luncheon I described above, I ran into the Associate Administrator for Aeronautics at NASA and asked him if we could talk. The result was my leaving FAA to go to NASA as the Director of Aviation Safety Research. From there I was chosen to be on the safety staff of the Presidential White House Commission on Aviation Safety and Security. That led to my being selected to be detailed to the White House as the Senior Policy Advisor for the Aviation in the National Science and Technology Council in the Executive Office of the President.

I tell you this story because I truly believe that the Holy Spirit guided my life and my career. At the low point, when I felt all was over, an entirely new set of opportunities emerged that led me to experiences and contributions that I could not have dreamed of. At each step along the way, I felt the pull as I did when I was first guided into the FAA. Having my 23rd Psalm experience led to my transition to NASA, and then a series of airplane accidents brought me to the White House. I was the Aviation Advisor in the White House on September 11, 2001 when the planes hit the World Trade Center, Pentagon and field in Pennsylvania. I was honored to serve the nation in the White House at that important time.

As events in my life and career occur, I began to search for the context of what was happening. Perseverance and faith has kept me moving forward into the unknown. The hand of God through His spirit has leaded me to the point where I am writing this book.

The way that God communicates with us can be understood by observing what is happening in real life, the context of our life, within the context of the Bible. **It seems that unlikely coincidences are often God's way of telling us to listen to the Spirit and to search for what He wants us to do.**

This book was meant to be and you were meant to read it.

What is the issue concerning you? What is its context? What Bible verses come to mind? What is their context within the Bible? Pray and listen. What is your heart / the Holy Spirit saying within those contexts? What happens next in real life? Look back on God's work in your life. Learn to listen to what the Spirit says to you.

Chapter 10. Listening to God

The fundamental conclusions of this book that I would like you to consider are these:

1. Jesus Christ was crucified, rose from the dead, and sends his Holy Spirit to join with the spirit of those who love God.
2. The Jews believed that the messiah would come to save the nation of Israel, but instead, He came to save individuals throughout the generations.
3. God wants a personal relationship with you.
4. This is true even if you have had a bad past experience in a particular church or with people who are acting according to their religious belief.
5. This is true even if you have had problems in your life.
6. The Gospel of John and the 7 letters written by Paul are the closest thing to what Jesus directly told us, both while He was alive and after his resurrection. They are Jesus' eyewitnesses. (1Thes, Gal, 1&2Cor, Rom, Phil, Philem)
7. The Holy Spirit teaches us through a combination of Bible study, personal experiences, and prayer.
8. The Bible is the inspired word of God, is filled with parables, and is a means that the Spirit uses to guide you. There is no single interpretation that can be proven.
9. The Bible and the Spirit combine to guide your life if you will read, listen, and learn.
10. Paying attention to the context enhances insight into the Spirit's guidance in Bible study and personal experiences.

Listening to God is a skill that is developed over a lifetime. C.S. Lewis' essay "Meditation in a Tool Shed"[39] helps us to understand how to listen:

---

39 C.S. Lewis, Originally published in The Coventry Evening Telegraph (July 17, 1945); reprinted in God in the Dock (Eerdmans, 1970; 212-15).

> "I was standing today in the dark tool-shed. The sun was shining outside and through the crack at the top of the door there came a sunbeam. From where I stood that beam of light, with the specks of dust floating in it, was the most striking thing in the place. Everything else was almost pitch-black. I was seeing the beam, not seeing things by it.
> Then I moved, so that the beam fell on my eyes. Instantly the whole previous picture vanished. I saw no tool-shed, and (above all) no beam. Instead I saw, framed in the irregular cranny at the top of the door, green leaves moving on the branches of a tree outside and beyond that, 90 odd million miles away, the sun. Looking along the beam, and looking at the beam are very different experiences."

Reading scripture or observing an event in your life is like looking at the beam. Listening to the Spirit is like looking along the beam. You need to look both at and along the flow of events to understand what is happening and glimpse the meaning the Spirit has for you. This is true of both scripture reading and of observing life experiences.

I suggest that you begin listening to the Spirit by reading John's Gospel and Paul's original letters. Then read the entire NT. As you read, highlight passages that stand out to you. Read the words and feel the flow of events. The passages that stand out are the ones that will give you insight into your personal spiritual context and the things you need to explore deeper. Set aside a quiet time at the beginning or the end of each day to reflect on the flow of events that are happening in your life. Pray for insight from the Spirit. As troubles occur, search for God's meaning for you and persevere. He will answer you even if it is not in the way you are expecting. Remember:

> Hebrews 12:6-11 (NASB)
> 6 FOR THOSE WHOM THE LORD LOVES HE DISCIPLINES, AND HE SCOURGES EVERY SON WHOM HE RECEIVES."
> 7 It is for discipline that you endure; God deals with you as with sons; for what son is there whom his father does not discipline?
> 8 But if you are without discipline, of which all have become partakers, then you are illegitimate children and not sons.
> 9 Furthermore, we had earthly fathers to discipline us, and we respected them; shall we not much rather be subject to the Father of spirits, and live?
> 10 For they disciplined us for a short time as seemed best to them, but He disciplines us for our good, so that we may share His holiness.
> 11 All discipline for the moment seems not to be joyful, but sorrowful; yet to those who have been trained by it, afterwards it yields the peaceful fruit of righteousness.

I have always felt that when problems occur, it is best to discover what God is trying to tell me, and others around me, and take corrective action as soon as possible. In this way, I hope to learn the lesson quickly and go on to the next training session. He has a bigger plan for you than you can imagine. The events in your life may not only be a lesson for you. God may be using you to help others receive God's message. During joyous times, recognize that they are a blessing from God. Remember, **we are all on our own path** to becoming sons of God, and it is not our place to judge others who are on their path.

Appendix 1 contains the chart that I used to organize this book. I hope that it is a useful summary for you to recall what has been presented in this book and as a tool to discuss how God speaks to you with others. Listening to God is active, not academic.

God first touched me when I was 15. I was walking home from high school wrestling practice in the dead of winter when I uttered a simple prayer under my breath. "Dear God, It sure is cold." At that moment I felt a wave of heat starting at my toes come over me to the tip of my head. I was invigorated. I ran home, burst into the kitchen where my mother was cooking dinner and announced that I was going to become a minister. It has been quite an adventure from that night to today, but the Spirit has finally brought me to the day when I am writing this book.

I am sure your experience will be different from mine, but I challenge you to think back to see when the Spirit has affected your life. Then look each day to see what He is doing in your life now. For one thing you are reading this book. The process I have just described will result in your walking with the Spirit in your life. That will result is a sense of assurance, freedom from fear, and the opportunity to experience Wonder, Love, Gratitude, and Exhilaration.

Chapter 11. The Bible is The Garden

But how do you know that it is the Holy Spirit that is guiding you and not your own desires or another spirit? The answer is most directly quoted from the Gospel of Matthew.

> Matthew 7:17-20 (NASB)
> 17 "So every good tree bears good fruit, but the bad tree bears bad fruit.
> 18 "A good tree cannot produce bad fruit, nor can a bad tree produce good fruit.
> 19 "Every tree that does not bear good fruit is cut down and thrown into the fire.
> 20 "So then, **you will know them by their fruits.**

This is the way to test your thoughts to see if they come from the Holy Spirit. Here is what Jesus says in the Gospel of John.

> John 15:1-5 (NASB)
> 1 "I am the true vine, and My Father is the vinedresser.
> 2 "Every branch in Me that does not bear fruit, He takes away; and **every branch that bears fruit, He prunes it so that it may bear more fruit.**
> 3 "You are already clean because of the word which I have spoken to you.
> 4 "Abide in Me, and I in you. As the branch cannot bear fruit of itself unless it abides in the vine, so neither can you unless you abide in Me.
> 5 "I am the vine, you are the branches; he who abides in Me and I in him, he bears much fruit, for apart from Me you can do nothing.

But what is good fruit? Peter and Paul answer this question.

1 Peter 3:8-9 (NASB)
8 To sum up, all of you be harmonious, sympathetic, brotherly, kindhearted, and humble in spirit;
9 not returning evil for evil or insult for insult, but giving a blessing instead; for you were called for the very purpose that you might inherit a blessing.

1 Thessalonians 5:14-24 (NASB)
14 We urge you, brethren, admonish the unruly, encourage the fainthearted, help the weak, be patient with everyone.
15 See that no one repays another with evil for evil, but always seek after that which is good for one another and for all people.
16 Rejoice always;
17 pray without ceasing;
18 in everything give thanks; for this is God's will for you in Christ Jesus.
19 Do not quench the Spirit;
20 do not despise prophetic utterances.
21 **But examine everything carefully; hold fast to that which is good;**
22 abstain from every form of evil.
23 Now may the God of peace Himself sanctify you entirely; and may your spirit and soul and body be preserved complete, without blame at the coming of our Lord Jesus Christ.
24 Faithful is He who calls you, and He also will bring it to pass.

What struck me as I wrote this book is that the Bible is the opposite of the creation story of the Garden of Eden. The Bible begins with God putting Adam and Eve into the Garden of Eden. He tells them that they are free to eat anything they want except from the tree of good and evil. When they disobey God, He

sends them out of the garden to keep them from eating of the tree of life.
The rest of the Bible is the opposite of the Adam and Eve parable. The Bible is the book that gives us knowledge of good and evil, and the cross of Christ is the tree of life. In the parable of the Bible, God wants us to know about good and evil, and to embrace the tree of life through the death of Christ Jesus. He has led us full circle. It took thousands of years for the Bible, the tree of good and evil, to be created through God's inspiration of perhaps millions of people who have touched it along the way. It took God sending his Son into the world to show us the way back to the garden and give us eternal life. The Bible tells us so.

> Genesis 2:8-9 (NASB)
> 8 The LORD God planted a garden toward the east, in Eden; and there He placed the man whom He had formed.
> 9 Out of the ground the LORD God caused to grow every tree that is pleasing to the sight and good for food; the tree of life also in the midst of the garden, and the tree of the knowledge of good and evil.

> 1 Corinthians 15:20-22 (NASB)
> 20 But now Christ has been raised from the dead, the **first fruits** of those who are asleep.
> 21 For since by a man came death, by a man also came the resurrection of the dead.
> 22 For as in Adam all die, so also in Christ all will be made alive.

Conclusion: As with the first man, Adam, who walked through the Garden of Eden and God talked to him, today God will talk to you through the Bible. God banished Adam and Eve from the Garden of Eden so that they would not eat the fruit of the Tree of Life. We now know that the tree of life is the cross of Jesus and through it we gain eternal life. We know this from the Bible. The

Bible is the garden in which the Holy Spirit works to produce good fruit in those who seek God.

Epilogue

John 1:12-13, "Yet to all who received him, to those who believed in his name, he gave the right to become children of God--, children born not of natural descent, nor of human decision or a husband's will, but born of God.

This book has been written so that you might see the big picture of how God speaks to you using the Bible and His Spirit. But, there is one question left unanswered. It is truly remarkable that God actually wants a personal relationship with each one of us. Why? In most other religions, the deity is the one to be sacrificed to. The deity does not sacrifice itself for its people as Jesus Christ did for us. Why does God want a relationship with you and me? I think I have the answer.....Love.

Love is the one thing that God cannot make us give Him because love is only real if another gives it freely and sincerely.

> 1 John 4:7-11 (NASB)
> 7 Beloved, let us love one another, for love is from God; and everyone who loves is born of God and knows God.
> 8 The one who does not love does not know God, for **God is love**.
> 9 By this the love of God was manifested in us, that God has sent His only begotten Son into the world so that we might live through Him.
> 10 In this is love, not that we loved God, but that He loved us and sent His Son to be the propitiation for our sins.
> 11 Beloved, if God so loved us, we also ought to love one another.

So if God wants a personal relationship with us because he loves us then what is God's ultimate purpose for our existence? I believe that Paul and John give us the answer.

Romans 8:14-31 (NASB)
14 **For all who are being led by the Spirit of God, these are sons of God.**
15 For you have not received a spirit of slavery leading to fear again, but you have received a spirit of adoption as sons by which we cry out, "**Abba!** Father!" (The same words Jesus used to refer to God)
16 The Spirit Himself testifies with our spirit that **we are children of God**,
17 and if children, heirs also, heirs of God and **fellow heirs with Christ**, **if indeed we suffer with Him** so that we may also be glorified with Him.
18 For I consider that the sufferings of this present time are not worthy to be compared with the glory that is to be revealed to us.
19 For the anxious longing of the creation waits eagerly for the revealing of the sons of God.
20 For the creation was subjected to futility, not willingly, but because of Him who subjected it, in hope
21 that the creation itself also will be set free from its slavery to corruption into the freedom of the glory of the children of God.
22 For we know that the whole creation groans and suffers the pains of childbirth together until now.
23 And not only this, but also **we ourselves, having the first fruits of the Spirit**, even we ourselves groan within ourselves, **waiting eagerly for our adoption as sons**, the redemption of our body.
24 For in hope we have been saved, but hope that is seen is not hope; for who hopes for what he already sees?
25 But if we hope for what we do not see, with **perseverance** we wait eagerly for it.
26 In the same way the Spirit also helps our weakness; for we do not know how to pray as we should, but the Spirit Himself intercedes for us with groanings too deep for words;

27 and He who searches the hearts knows what the mind
of the Spirit is, because He intercedes for the saints
according to the will of God.
28 And we know that **God causes all things to work
together for good to those who love God, to those
who are called according to His purpose.**
**29 For those whom He foreknew, He also predestined
to become conformed to the image of His Son, so that
He would be the firstborn among many brethren;**
**30 and these whom He predestined, He also called;
and these whom He called, He also justified; and
these whom He justified, He also glorified.**
31 What then shall we say to these things? If God is for
us, who is against us?

Revelation 1:5-6 (NASB)
5 and from Jesus Christ, the faithful witness, the firstborn
of the dead, and the ruler of the kings of the earth. To Him
who loves us and released us from our sins by His
blood—
6 and **He has made us to be a kingdom, priests to His
God and Father**—to Him be the glory and the dominion
forever and ever. Amen.

These passages tell us that our ultimate roll is to be God's sons, heirs to his kingdom, and his priests as Jesus is.

John agrees with Paul in the opening of his Gospel.

John 1:9-13 (NASB)
9 There was the true Light which, coming into the world,
enlightens every man.
10 He was in the world, and the world was made through
Him, and the world did not know Him.
11 He came to His own, and those who were His own did
not receive Him.

> 12 **But as many as received Him, to them He gave the right to become children of God, even to those who believe in His name,**
> **13 who were born, not of blood nor of the will of the flesh nor of the will of man, but of God.**

Notice that John says that God gives the **right**, not the immediate result to those "not born of Blood". If we read down to Chapter 3 of John's Gospel an amazing idea emerges.

> John 3:4-8 (NASB)
> 4 Nicodemus *said to Him, "How can a man be born when he is old? He cannot enter a second time into his mother's womb and be born, can he?"
> 5 Jesus answered, "Truly, truly, I say to you, unless one is born of water and the Spirit he cannot enter into the kingdom of God.
> 6 "**That which is born of the flesh is flesh, and that which is born of the Spirit is spirit.**
> 7 "Do not be amazed that I said to you, **'You must be born again.'**
> 8 "The wind blows where it wishes and you hear the sound of it, but do not know where it comes from and where it is going; so is everyone who is born of the Spirit."

What is God's ultimate purpose? I propose the following: **God is spirit and He is creating a spiritual family for Himself from us here on earth.** While being "born again" has taken on the meaning of receiving Christ into our lives by some of today's churches, I don't think that is all that John is talking about. Let me explain.

John 3:5 says: "That which is born of the **flesh is flesh**, and that which is born of the **Spirit is spirit**". This is specifically saying that we, as individuals, must at some point be born other than in the flesh if we are to be His sons. Yes we must receive Christ as our savior, but this is the beginning not the end of the process. Because we do not have any firsthand knowledge of how we might be actually born of the spirit, we have interpreted it to be something that we can understand, receiving Christ into our lives. However, Nicodemus didn't understand it this way. He was certainly amazed at the thought of a second birth. However, there is an interpretation that would explain this that is consistent with Biblical text. Once seen the story line of the Bible changes.

Genesis says that God starts by creating the Heavens and the Earth. Then He creates men and women. We are told that men and women are created in God's image. (We have a spirit and a physical form like Jesus). However, we not only resemble Him, but He has also created us so that we can reproduce. The man's sperm impregnates the woman's egg and another human being is formed. That human being is "begotten" by two humans and has it's own spirit from its mother and father. This human spirit has the free will to seek God or turn away.

The hypothesis is this: the human spirit is the spiritual egg of a spiritual person. If a person seeks God, God impregnates the human spirit with the Holy Spirit, (i.e. born again by receiving Christ as we currently think today). Under this hypothesis, the two spirits (God's and the individual's) become a spiritual embryo and eventually we are actually born as a spiritual person, a son of God, with no physical body. If you think back to the story of the Virgin Mary we see this described in part in the birth of Jesus.

> Luke 1:35 (NASB)
> 35 And the angel said to her, "The Holy Spirit will come upon you, and the power of the Most High will overshadow you; therefore the child to be born will be holy; he will be called Son of God."

Because Mary was a virgin, Jesus was not born with a human spirit. He had the full Spirit of God within his human body. He was not an embryo as we are. He has both a physical and spiritual body.

Under this hypothesis, the spiritual person is born after the physical death of the physical person. We are literally born again in the spirit as God's son. This would be you with all your memories, skills and abilities yet in a spiritual body that is recognizable as you. As Paul says in Romans Verse 23 above, we only have the first fruits of the Spirit now and are awaiting our full adoption / birth as sons. That makes us spiritual embryos throughout our physical lifetime. The human spirit has the choice to love God or turn away. If the choice is to love God, then the Holy Spirit comes upon us as he did with Mary and creates a spiritual embryo comprised of our human spirit and the Holy Spirit. In this way, God creates a true loving spiritual family comprised of "sons of God" that truly love him. And, Jesus is the first fruit, so we will be like him. Here is what Paul has to say:

> 1 Corinthians 15:42-53 (NASB)
> 42 **So also is the resurrection of the dead. It is sown a perishable body, it is raised an imperishable body;**
> 43 it is sown in dishonor, it is raised in glory; it is sown in weakness, it is raised in power;
> 44 it is sown a natural body, it is raised a spiritual body. If there is a natural body, there is also a spiritual body.
> 45 So also it is written, "The first MAN, Adam, BECAME A LIVING SOUL." The last Adam became a life-giving spirit.
> 46 However, the spiritual is not first, but the natural; then the spiritual.
> 47 The first man is from the earth, earthy; the second man is from heaven.
> 48 As is the earthy, so also are those who are earthy; and as is the heavenly, so also are those who are heavenly.

49 Just as we have borne the image of the earthy, we will also bear the image of the heavenly.
50 Now I say this, brethren, that flesh and blood cannot inherit the kingdom of God; nor does the perishable inherit the imperishable.
51 Behold, I tell you a mystery; we will not all sleep, but we will all be changed,
52 in a moment, in the twinkling of an eye, at the last trumpet; for the trumpet will sound, and the dead will be raised imperishable, and we will be changed.
53 For this perishable must put on the imperishable, and this mortal must put on immortality.

We will be born again as sons of God.

If this theory is correct, the real purpose of our lives is to learn to love God and others, grow in His love, and develop the knowledge, skills, and abilities that we will need to play a role as His sons in His kingdom. What a fantastic future! We will be part of God's family, ruling the universe that He has created with, through, and for love. Try reading the Bible with this idea in mind. It is amazing how difficult passages and events in our lives become more understandable and bearable.

God Bless You Brother. For whether we are male or female, we will all be sons of God.

Appendix 1: God's Communications Model

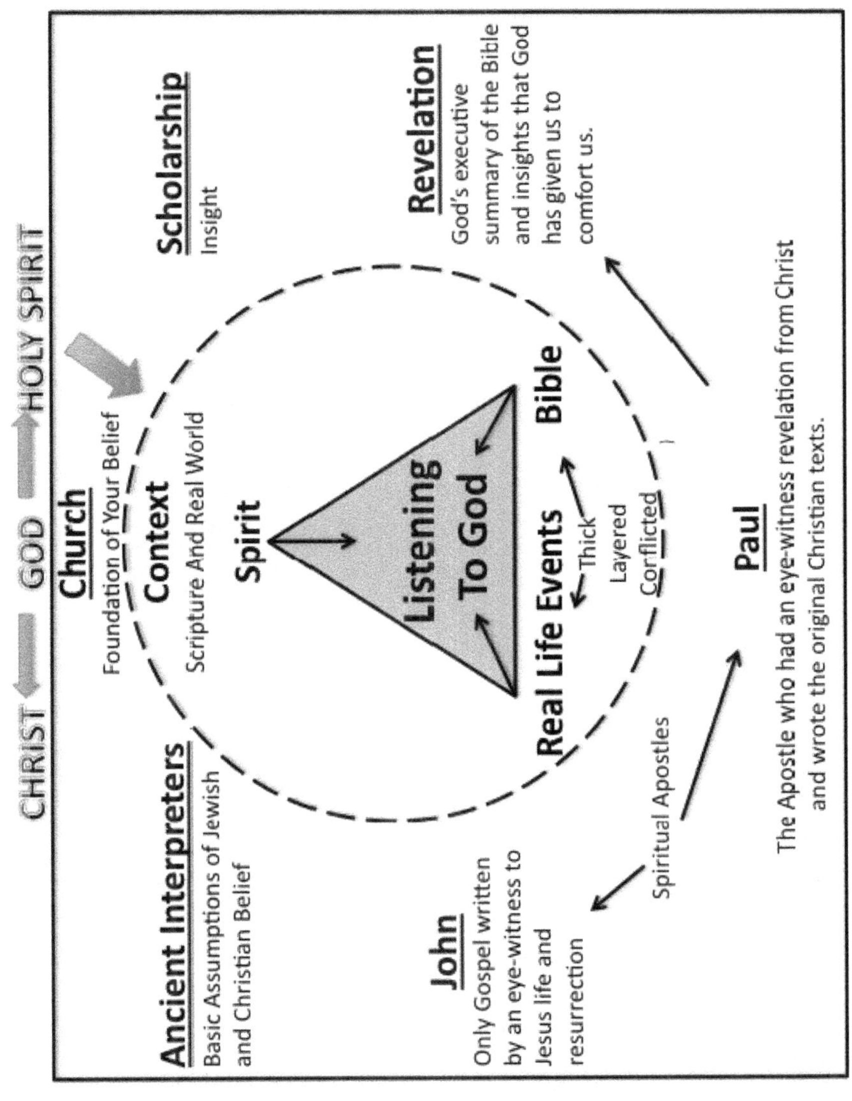

Appendix 2: Another Way God Speaks To Us – Mathematics

Another Way God Speaks To Us – Mathematics

Then God said, "Let there be light"; and there was light.[40]

There has long been a tension between science and religion. Science is excellent at creating a mathematical hypothesis and through the "scientific method" scientists use measurement and analysis to prove or disprove the hypothesis. The most fundamental tool that science uses is mathematics. Mathematics can describe anything that exists from the cosmos to subatomic particles. By extending mathematical formulas beyond what is known, the unknown can be predicted. Then, scientists find ways to test the mathematical hypothesis thereby advance science and human knowledge. Once proven, the hypothesis becomes a theory, and if the theory works every time it is applied, we call it a law of physics. This process has lead to some amazing conclusions in the last 100 years. While some scientists may not realize it, mathematics is a way that God speaks to us.

Isaac Newton discovered the basic formulas that describe what our senses show us. His monograph Philosophiæ Naturalis Principia Mathematica, published in 1687, lays the foundations for most classical physics. In this work, Newton described universal gravitation and the three laws of motion, which dominated the scientific view of the physical universe until the 20th century.

By 1901 scientists and mathematicians started to unlock a deeper reality.

---

[40] *Genesis 1:3 (NASB)*

A German physicist, Max Planck, discovered quantum physics from his research on the properties of light. He is regarded as the founder of quantum theory, for which he received the Nobel Prize in Physics in 1918.

Around the same time, Albert Einstein[41] a German-born theoretical physicist developed the theory of general relativity, effecting a revolution in physics. For this achievement, Einstein is often regarded as the father of modern physics and one of the most prolific intellects. Near the beginning of his career, Einstein thought that Newtonian mechanics was no longer enough to reconcile the laws of classical mechanics with the laws of the electromagnetic field. This led to the development of his special theory of relativity. He realized, however, that the principle of relativity could also be extended to gravitational fields, and with his subsequent theory of gravitation in 1916, he published a paper on the general theory of relativity. He continued to deal with problems of statistical mechanics and quantum theory, which led to his explanations of particle theory and the motion of molecules. He also investigated the thermal properties of light that laid the foundation of the photon theory of light. In 1917, Einstein applied the general theory of relativity to model the structure of the universe as a whole.

From this beginning, scientists and mathematicians have made immense strides in defining the basic nature of the universe and perhaps beyond. As Professor Neil deGrasse Tyson says, "What ever we determine to be true (about the universe) may only be a subset of a larger truth."[42] I believe that this statement is prophetic in two ways. One is the direct meaning of the quote. Scientists continue to find deeper truths using the concepts of relativity theory, that describe the actions of very big things like

---

[41] history.http://en.wikipedia.org/wiki/Albert_Einstein - cite_note-1

[42] Professor Neil deGrasse Tyson, The Great Courses, *My Favorite Universe*, lecture 8, *In Defense of the Big Bang*, © The Teaching Company, 2003

solar systems and the cosmos, quantum mechanics, that describe very small things like atoms and their sub particles, and string theory, that attempts to merge the two. The other meaning of larger truth is that perhaps we are discovering the mind of God.

There is no way that I can present the history, detail the findings, or quote the sources of math and physics discoveries over the past 100 years in this appendix. So, I will suggest three books for you to read to gain insights into the astounding things that have been discovered; *Death by Black Hole*[43], by Neil deGrasse Tyson, *The Elegant Universe*[44] by Brian Greene, and *The Dreams That Stuff is Made Of,*[45] by Stephen Hawking. What I will describe are some of the conclusions that have come from these discoveries.[46] These discoveries make mathematical sense, not intuitive sense, yet they have been proven to agree with actual measurements. These conclusions have been tested and verified billions of different ways to the point that they prove that relativity theory and quantum mechanics are more than mathematical theories, they are more accurate than the laws that Newton discovered.

1. Speed is a measure of space that something travels over time in relation to something else.
2. Space and time adjust to result in the same speed of light (670,616,629 mph). As something accelerates through

---

43 *Death By Black Hole and Other Cosmic Quandries, Neil deGrasse Tyson, W.W. Norton & Company 2007.*

44 *The Elegant Universe,* Brian Greene, Vintage Books, 2000-2003

45 *The Dreams That Stuff Is Made Of,* Stephen Hawking, Running Press, Philadelphia, 2011

46 Background and descriptions of many of these conclusions can be seen in the PBS NOVA series *The Fabric of the Cosmos*, 2011, Brian Green, I cannot possibly footnote all the books and sources for these conclusions.

space, time for it slows down to the point that time for it stands still at the speed of light.

3. Space is not empty. It has properties. It contains the three dimensions that we see, a fourth dimension, time, and perhaps six or seven more that we cannot see. At the subatomic scale, "empty" space is flooded with activity that can force objects to move.
4. Time is not constant. It is dependent on motion and gravity.
5. Gravity is the shape of space-time. The heavier the mass the more the fabric of space-time is warped like a ball on a sheet of rubber.
6. Light has characteristics of a particle (photon) and a wave. It is an electromagnetic wave that never stops and only appears as a particle in a given location when it is measured. Otherwise there is only a probability that it is in a given location.
7. Understanding the nature of light has lead to the discovery of quantum mechanics that has led to the understanding that all matter and energy have similar fundamental properties.
8. Mass can be converted to energy and visa versa. Energy = Mass x Speed of Light x the speed of light again. At the smallest level, energy and mass are one.
9. The light from each element in the periodic table of elements has a specific spectrum signature when it is heated enough to give off light.
10. The frequency of light, sound and other electromagnetic waves shifts lower when an object is coming toward you (blue shift) and higher when it is moving away (red shift). This is called the Doppler shift and is a way to measure speed.
11. A body in space moving away from the earth is looking back in time if it looks back at the earth. If it is moving toward the earth it is looking to earth's future.

12. Now is a slice of space-time that is unique to the observer. Time is different for everyone based on his or her movement in space.
13. From a mathematical prospective the past and future already exist.
14. All bodies in the universe are moving away from earth and their speed is accelerating as it would resulting from an explosion.
15. By reversing the observed mass and motion, scientists conclude that the universe we know resulted from the "Big Bang" 13.7 billion years ago. Our universe had a beginning and is expanding at an increasing rate of speed.
16. The Big Bang is what moves time that we observe forward. All we ever experience is now. The flow of time may be an illusion.
17. Black holes are at the center of galaxies and stars that have collapsed. Their centers are of infinite densities and they get bigger the more mass that enters them. The outer edge of a black hole is called the "event horizon". Inside the event horizon, the gravitational pull is so great that light cannot escape.
18. Information about all that has gone into a black hole is stored on the boundary layer rim of the black hole as a two dimensional image. This could make a hologram of all that has gone into the black hole.
19. There is a tendency of everything in the universe to move from order to disorder, called entropy.
20. Energy comes in distinct chunks that cannot be subdivided called quanta.
21. Electrons are probability waves. You can never predict where they are.
22. All matter in the universe is made up of atomic and subatomic particles that are ruled by probability not certainty.

23. Two particles can become linked if they are close in proximity and have similar properties. This is called entanglement. Once entangled, they stay entangled no matter how far they are separated. Measuring one will result in an effect on the other as if the space between them doesn't exist. This may be faster than the speed of light. While this violates our space-time thinking, it has been proven to be true. Einstein called this "spooky action".
24. The fundamental nature of reality at the deepest level is determined by chance. All is uncertain at the smallest particle level until something is measured. The moment you observe a particle the uncertainty disappears.[47]

Where is this leading scientists and mathematicians? Here is a list of areas that have not been proven, but scientists are actively investigating.

1. The ultimate reality could be that our three dimensional universe is a projection of a two-dimensional surface from the boundary layer of a Black Hole at the edge of the universe.
2. There probably are multiple universes.
3. The Big Bang may result from an initial condition prior to the Big Bang where gravity reverses itself, called inflation. According to the theory, Inflation doesn't end everywhere at once. When it does, a universe is formed through a big bang.
4. At the smallest level, the universes are probably composed of vibrating energy strings that take on different properties depending on their vibration. These strings vibrate through 10 dimensions.

---

[47] Neils Bohr

5. Atoms cannot pass from one universe to another, but gravity may transcend universes. We might be able to detect the other universes from ripples in gravity.

Where does this lead us? What is the math telling us? The greatest finding is that there are layers of reality beyond what we see and feel with our five senses. The reason we don't recognize the things I just described in our daily lives is because they are either operating at such a large scale or a small scale that they are not apparent. However, they are real.

Science set out to discover the answers to what everything is made of and how it works. Some scientists feel that God is an invention of man's mind to explain things that are unknown. Therefore, if science could discover what is unknown, then God and faith are not needed to explain reality.

Are we a projection of a multi-dimensional hologram? Are we the random result of energy strings in a multi-universe that is so large that we all just happen to exist as we do? What ever we may discover to be true, there is always the question of how did this come to be? Saint Thomas Aquinas had an answer in the thirteenth century. He essentially said in his contingency argument; nothing has been discovered that carries with it the reason for its own existence except God who identifies Himself as "I Am".[48]

What scientists have in fact discovered is that what is real in this universe and beyond is only a probability not an answer. There are firm laws that describe everything, but what they describe is not definitive. At the smallest level, there is only a probability that matter exists at any point in space-time.

I can't help thinking about a quote from the Apostle John.

---

[48] *Summa Theologica* written 1265–1274

1 John 1:5 (NASB)
5 This is the message we have heard from Him and announce to you, that **God is Light**, and in Him there is no darkness at all.

This quote and the opening of Genesis both illustrate that part of God's nature is characterized by light. Isn't it fascinating that understanding light has resulted in our understanding of the building blocks of the universe through the discovery of quantum mechanics. And, because light travels forever and time stands still at the speed of light, light is immortal as is God. I believe that what math and science has discovered are the basic building blocks that God has used in creation. God's language is mathematics. The super string of quantum mechanics is the DNA of physical creation, where biological DNA is the basic building block of life. God is consistent in using fundamental codes and equations to create the universe and life within it.

A second quote from John also comes to mind.
1 John 4:8 (NASB)
8 The one who does not love does not know God, for **God is love**.

Another and most important part of God's nature is love. God is greater than what can be defined by man, but he has given us the ability to know him personally through the Word and Spirit as I have described in this book. He has done this because of his love for us. He has also allowed us to know how He has formed His creation through mathematics, and science derived from the understanding of what light is.

In chapter 3 of this book, I wrote the following about the nature of the Bible:

> While scholar's search for many years has been to find a definitive answer to questions posed by Biblical text, it has

> actually shown that no single answer exists. In fact, Biblical scholarship has shown the Bible to be like the atom, the deeper you look the more you find that there is no mass there, only energy and a probability of mass. Perhaps that says something about God.

Scientists are finding out that their search to understand the universe is like the Biblical scholars search to understand the Bible. Biblical scholars, mathematicians, and scientists have searched for certainty, but it does not exist. This is a common thread between physics and Bible scholarship. Yet, in spite of this bottom line of probability and uncertainty, the Bible does exist, and our Universe exists. I contend that, it is God who turns probability into reality. What man sees as uncertainty, is God's flexibility. God used basic codes and laws to create everything, but didn't constrain His ability to act in His creation by building His creation on uncertainty... God built our universe on uncertainty so that he could act in it without violating His own laws of creation. This may be how God makes things happen in your life, and how miracles occur without violating the laws of physics.

Again, as Professor Neil deGrasse Tyson says, "What ever we determine to be true (about the universe) may only be a subset of a larger truth."[49] Without knowing it, scientists may be discovering that God is all that is real and our best science and math point to that reality.

---

[49] Professor Neil deGrasse Tyson, The Great Courses, *My Favorite Universe*, lecture 8, *In Defense of the Big Bang*, © The Teaching Company, 2003

CPSIA information can be obtained
at www.ICGtesting.com
Printed in the USA
FSHW021412031221
86521FS

9 781621 416883